D1616852

ADVANCE PRAISE FOR

# *A Hinge of History*

"One of our great statesmen leads an exploration of the most important issues that our nation faces. This is a rich menu for our future."

—**Joseph S. Nye Jr.**, Harvard University, author of *Do Morals Matter? Presidents and Foreign Policy from FDR to Trump*

"Improving how our government works will restore confidence in our institutions during this critical time in our history. Secretary Shultz has once again made an important civic contribution."

—**Jeb Bush**, former governor of Florida

"The Governance in an Emerging New World project is vital to our time. We need these sorts of constructive debates to respond thoughtfully in the midst of unprecedented technological, social, political, and economic change."

—**Ralph D. Semmel**, director of the Johns Hopkins Applied Physics Laboratory

"In a world awash in change, governing effectively demands more breadth and depth of understanding than we've achieved historically. This clear-eyed, smartly presented study enables a unique grasp of what is at stake, without myth or delusion, and provides the intellectual tools for sorting out the path ahead."

—**James Mattis**, general, US Marines (ret.) and 26th secretary of defense

"A strategic analysis of the disparate and sometimes conflicting factors shaping the future that should inform policy makers and students for thinking beyond narrow disciplinary silos. It bridges the sometimes perilous knife-edge of science, technology, and security in ways few other volumes have."

—**Margaret Kosal**, professor of international affairs, Georgia Tech

"An amazing tour de force through the most challenging governance issues of our times. Its analytical depth and the breadth of its thematic and geographical scope make it essential reading for policy makers and observers around the globe."

—**Jens Suedekum**, professor of international economics, Düsseldorf Institute for Competition Economics

"The world grapples with a pandemic, demographic shifts have remade entire regions, and authoritarianism has surged. Amid this crisis, this volume reminds us that restoring global leadership requires two things: reflection and collaboration. Global security, human rights, and the spirit of democracy all demand it."

—**H. E. Hicham Alaoui**, director of the Hicham Alaoui Foundation for Social Science Research on Northern Africa and the Middle East

"Timely and important ideas for tackling our toughest global challenges. *A Hinge of History* is a valuable resource for leaders and citizens who recognize that facts, not ideology, must shape our conclusions and our path forward."

—**Sam Nunn**, former US senator and chair of the Senate Armed Services Committee, cochair of the Nuclear Threat Initiative

"Few in our nation's history have brought to public service the breadth of talent and expertise that George P. Shultz has. Here, he and Jim Timbie share the results gained in their ambitious project to explore how changing demographics, technology, education, and the information revolution will shape the world ahead." —**Karen Tumulty**, *Washington Post*

"Governance everywhere is made more difficult by disruptive forces. Existing institutions and their leaders are struggling to know how best to act and adjust. Shultz shows how careful study and leadership can once again address today's unfolding 'emerging new world' problems."

—**Van Ton-Quinlivan**, CEO, Futuro Health, and former executive vice chancellor, California Community Colleges

"A dangerous consequence of the current pandemic is that it is fostering isolationism among nations at a time when international cooperation is needed the most: an effective global governance is one of the keys for a prompt recovery." —**Pedro Aspe**, former treasurer of the republic, Mexico

"*A Hinge of History* can serve as a blueprint for the next administration in Washington to begin the process of restoring American leadership with a renewed sense of purpose to tackle the biggest issues in global and human development." —**James Hollifield**, professor of political science and director of the Tower Center, Southern Methodist University

# A Hinge of History

# A Hinge of History

## GOVERNANCE IN AN EMERGING NEW WORLD

George P. Shultz and James Timbie

HOOVER INSTITUTION PRESS
STANFORD UNIVERSITY | STANFORD, CALIFORNIA

*With its eminent scholars and world-renowned library and archives, the Hoover Institution seeks to improve the human condition by advancing ideas that promote economic opportunity and prosperity, while securing and safeguarding peace for America and all mankind. The views expressed in its publications are entirely those of the authors and do not necessarily reflect the views of the staff, officers, or Board of Overseers of the Hoover Institution.*

*hoover.org*

**Hoover Institution Press Publication No. 717**

Hoover Institution at Leland Stanford Junior University,
Stanford, California 94305-6003



First printing 2020
27 26 25 24 23 22 21 20 9 8 7 6 5 4 3 2 1

Manufactured in the United States of America
Printed on acid-free archival-quality paper

Library of Congress Control Number: 2020943759

ISBN: 978-0-8179-2434-8 (cloth)
ISBN: 978-0-8179-2436-2 (epub)
ISBN: 978-0-8179-2437-9 (mobi)
ISBN: 978-0-8179-2438-6 (PDF)

# Contents

# List of Figures

# Preface

On April 8, 2019, we gathered around the circular table in the Annenberg Conference Room at the Hoover Institution for another discussion from our research project Governance in an Emerging New World. This session was led by Dr. Lucy Shapiro, a professor of biology at the Stanford University School of Medicine.

One of the featured speakers, Dr. Milana Boukhman Trounce, a professor and emergency medicine physician at Stanford, presented a paper on "Potential Pandemics." She described in detail in her paper how "the threat of infectious disease is making a comeback. Unfortunately, at this point we are ill equipped to deal with a number of scenarios, particularly those involving large-scale infectious disease outbreaks—pandemics." She explained how human activities, including increased contact between humans and wild animals and global transportation networks, have increased the threat of new infectious diseases. She explained why drugs for treatment of such a new disease and vaccines to prevent its spread would not be available in time to prevent a public-health crisis. The principal countermeasures would be the same public-health measures that have been used for centuries—isolation and quarantine. We learned that the word *quarantine* is derived from an Italian term for the forty days that all ships were required to be isolated before passengers and crew could go ashore during the Black Death plague epidemic in the fourteenth century. We learned about "social-distancing" measures, such as closing schools, public gatherings, businesses, and transport, and the possibility that internet commerce might facilitate implementation of isolation and quarantine. We learned that rapid diagnostic testing, if available, might help a lot. Later in the day, we heard from Stephen Quake, a professor of bioengineering at Stanford, about the potential of modern gene-sequencing technology to rapidly recognize the causes of infectious disease outbreaks.

In response, Dr. Shapiro observed, "If you don't push the boundaries of understanding this world that we are living in . . . without new kinds of understandings of how living beings, living organisms, can survive changes in their environment—we are being, if not short-sighted, then we are being criminal."

A year later, our authors' meeting to discuss the editing of this manuscript with our publisher was held not around the conference room table of the Hoover Institution Press office on Stanford campus, but virtually over Zoom, each team member in his or her spare bedroom—three months after a novel coronavirus became capable of infecting and spreading among humans in Wuhan. The potential threat was known, but we were not prepared for it when it became real. In our project we have learned of other potential threats, including large-scale migration, global warming, and nuclear proliferation, not to mention the potential for infectious diseases far more lethal than COVID-19. Knowledge can lead to assessment of risk and appropriate preparations. We hope this volume will contribute to such outcomes.

The virus shows that we are part of an interconnected world. And a pandemic is a clear example of a problem whose solution would benefit greatly from international cooperation and US leadership. The global response to COVID-19 could have been more effective with better international cooperation—and better US leadership. This is a recurring theme of our project. International cooperation is part of the solution to the transformational challenges before us, including advancing technology, changing demographics, migration, global warming, nuclear proliferation, and health. We hope this volume will support US leadership at this crucial time in our history.

# Acknowledgments

This book is one result of a team effort on a wide-ranging project that took place over two years at the Hoover Institution and that drew upon support from across the Institution's family of scholars.

We would in particular like to acknowledge project team members David Fedor and James Cunningham for their work in organizing project discussion sessions and for their hand in the shared reflections (and this manuscript), which flowed from them. We also thank Hoover Annenberg distinguished fellows and visiting fellows Adele Hayutin, Jim Hoagland, Ambassador David Mulford, and Admiral James O. Ellis Jr. (USN, ret.) for contributing their observations on global demographics, on Europe, on India, and on national security; Ambassador George Moose of the United States Institute for Peace, for his keen observations on Africa; and the fine staff members of Hoover events (Janet Chang Smith), communications (Shana Farley and Rachel Moltz), and Hoover Institution Press (Barbara Arellano, Danica Michels Hodge, Marshall Blanchard, and Alison Law, with freelance editor Mike Iveson) for their continued service to promote these and other ideas defining a free society.

Gifts from the individual supporters of the Hoover Institution, including John A. and Cynthia Fry Gunn, are what enabled all of that to take place. We are grateful to them for their continued animation of Hoover's policy research mission.

Throughout this book we have referred to a series of rich roundtables and public panel discussions hosted by the Hoover Institution on the Stanford University campus. These sessions underlie the development of our thinking on the "hinge of history," and each was backed by a series of original essays written for the occasion. These ten-to-fifteen-page essays are rich, provocative, and utterly approachable. Our editorial guidance to their authors was not to do original research (though many did). Rather, it was simply to sit down and draw upon their considerable existing expertise and experience in a certain region, or in a certain field, and to think about how the themes of the emerging new world—advancing technology, changing demographics, governance—were affecting that corner of the world. Those papers were compiled and distributed as booklets to roundtable discussants and panel audiences in advance of each session—copies of which proved far more popular than the free pizza that was also on offer to participating students. Together, these essays offer a variety of perspectives on a single theme, which is a necessary ingredient given that the questions and solutions we were concerned with are not black and white. Our goal in telling you all of this now is to try to convince you to read them. If any chapter or claim in this book has piqued your curiosity, there is more for you yet in our virtual archive, which you may access digitally at no cost by visiting hoover.org/governanceproject.

GEORGE P. SHULTZ and JAMES TIMBIE
*Hoover Institution, Stanford, California, May 2020*

# PART I

# Introduction to the Hinge of History

At the end of World War II, we were led by some gifted, tough-minded people with names like Dean Acheson and George C. Marshall and Harry S. Truman. When they looked back, what did they see? They saw two world wars, the United States drawn into both of them. They saw that the first one was settled in rather vindictive terms that helped lead to the second. They saw that fifty-one million people had been killed in that Second World War. They saw the Holocaust. They saw the Great Depression, which was a global event. They saw the protectionism and currency manipulation that had aggravated it. They said to themselves, "What a crummy world," and in contrast to the time at the end of First World War, when we walked away and refused to join the League of Nations, they also said, "and we are part of it whether we like it or not."

They set out to produce something better. Unlike the vindictive peace after World War I, they would try to help to rebuild Japan and Germany with the Marshall Plan and help them to become healthy democracies. And it's important to understand that the Marshall Plan was not the United States telling people what to do. Rather, it said—including to our former enemies—"Your economy is a mess, what do you think should be done about it? Let's talk about it, and probably you need some aid, so here it is." The idea was to have the United States provide resources with a framework, and to draw countries into doing something about their economies that would work.

Then there was the Bretton Woods system, in which forty-four countries were represented from the start, so it was not just a US effort. It was the British economist John Maynard Keynes, for example, who carried much of the water intellectually for the creation of institutions like the International Monetary Fund (IMF) through those global financial negotiations, which started among the Allied powers in 1944, even before the war had ended.

Others contributed substantively as well. But it relied on US leadership. Out of that came not just the IMF to deal with currency issues, but also the International Bank for Reconstruction and Development, now the World Bank, to deal with development issues and then the successive rounds of the General Agreement on Tariffs and Trade (GATT), which morphed into the World Trade Organization. Together they provided the framework for a long-term international economic regime. This foundation of international institutions would also help each of its members better face the world's new challenges through structures that ensured a degree of compromise and cooperation.

And those challenges did come, through the Cold War, for example, from which the creation of NATO was another shared response. In the early 1980s, the Soviets had intermediate-range nuclear weapons aimed at allies across Europe, and at Japan and China too. Their goal in doing so was to try to separate us as Americans from our allies and partners, asking if we would risk retaliation from their ICBMs for challenging their show of force. So, we made a deal with NATO to negotiate with the Soviets about it, and if we couldn't come to an agreement, then we would deploy our own intermediate-range missiles as a response. At the time, President Reagan knew he was negotiating with the European publics as much as he was negotiating with the Soviet Union, given the risk to them of potentially hosting nuclear missiles. That we continued with these negotiations even following the 1983 Soviet downing of Korean Airlines flight 007 showed the European public that we were doing so seriously and thoughtfully. Eventually, lacking progress, we deployed new cruise missiles in the United Kingdom and Italy with their cooperation, followed by the big deal: ballistic missiles in Germany. The Soviets walked out of negotiations, which rallied NATO, and finally, the intermediate-range Pershing missiles were deployed. Working together, NATO had turned the tide in the Cold War. A few months later, we were able to go to the president and say, "At four different capitals in Europe, a Soviet diplomat has come up to ours and said the same thing—that is, if Soviet foreign minister Andrei Gromyko is invited to Washington when he comes to the UN General Assembly meeting in September of 1984, he would accept." In other words, the Soviets blinked. Everything restarted.

By the time the Cold War ended, there had been created a pervading security and economic commons in the world from which everybody benefited. It's important to understand that there wasn't some grand design that led to that commons. Instead, it was created as we went along, with a lot of leadership from the United States, which was to our advantage. But the reason that system persisted even as allies and old foes regained their own strengths was that it had not been done purely in the service of our selfish national interests—the rest of the world, too, could benefit from being a part of it.

Over this time the globe saw truly great achievements. At the end of World War II, over two-thirds of the world's population lived in extreme poverty—that figure today is less than one in ten. The decline in the global illiteracy rate has followed the same path, alongside broader and deeper education. With improvements in health and nutrition and sanitation, longevity has risen by half. In 1950, one out of five of children around the world died before their fifth birthday from disease or malnutrition; that has improved to one in twenty. And democracy has spread: less than one-tenth of the world's population lived in free societies at the end of the war; today more than half do.[1] US economic and military resources helped.

But our world today is awash in change. That commons, built up piece by piece, is eroding. And with governance in disarray everywhere, there is hardly any place you can look in the world and say, "There is an island of stability and prosperity." Somehow, we are turning our backs on all of those constructive things we had built up together. We in the United States are facing another inflection point—much like the one Acheson and Marshall and Truman did in the 1940s, even if they didn't know it at the time. This is a hinge of history.

Sharp changes are afoot throughout the globe. The security and economic commons, already under threat, is being hit by new and important forces. Global demographics are shifting, technology both civilian and military is advancing at unprecedented rates, and these changes are being felt everywhere. Our world is most presently gripped by a contagious pandemic that has swept through our economies like a massive tidal wave. Other risks could overwhelm us again too—new threats like climate change, and nuclear weapons, which are again on the rise.

How should we develop strategies to deal with this emerging new world? We can begin by understanding it.

First, there is the changing composition of the world population, which will have a profound impact on societies. Essentially all developed countries are experiencing falling fertility and increasing life expectancy (and by fertility we mean not one's ability to reproduce, but rather the total number of children that are expected to be born to each woman in a country—akin to family size). These are the results of prosperity and health improvements, alongside women's education and cultural changes. As working-age populations shrink and pensions and care costs for an aging society rise, it becomes harder for governments to afford other productive investments.

At the same time, high fertility rates in Africa and South Asia are causing both working-age and total populations to grow, but that growth outpaces economic performance. And alongside a changing climate, these parts of the world already face growing impacts from natural disasters, human and agricultural diseases, and other resource constraints.

Taken together, we are seeing a global movement of peoples, matching the transformative movement of goods and of capital in recent decades—and encouraging a populist turn in world politics.

Second is automation and artificial intelligence (AI). In the last century, machines performed as instructed, and that "third industrial revolution" completely changed patterns of work, notably in manufacturing. But machines can now be designed to learn from experience, by trial and error. Technology will improve productivity, but workplace disruption will accelerate—and will be felt not only by call-center responders and truck drivers but also by accountants, by radiologists and lawyers, even by computer programmers.

All history displays this process of change. What is different today is the speed. In the early twentieth century, American farm workers fell from half the population to less than 5 percent alongside the mechanization of agriculture.[2] Our K–12 education systems helped to navigate this disruption by making sure the next generation could grow up capable of leaving the farm and becoming productive urban workers. With the speed of artificial intelligence, it's not just the children of displaced workers but the workers themselves who will need a fresh start.

Underlying this task is the reality that there are now roughly seven million "unfilled jobs" in America.[3] Filling them and transitioning workers displaced by advancing technology to new jobs will test both education (particularly K–12, where the United States continues to fall behind, condemning whole swaths of our young people to less satisfactory lives) and the flexibility of workers to pursue new occupations throughout their lives.

The third trend is fundamental change in the technological means of production, which allows goods to be produced near where they will be used and may unsettle the international order. In developed countries, we can now clearly see a novel shift toward dematerialization across a variety of sectors:

economic growth becoming decoupled from raw material input use. This happens as value generation shifts toward services and knowledge. But it also occurs as competition and new technologies drive production and distribution systems to ever greater efficiencies. More sophisticated use of robotics alongside human colleagues, plus additive manufacturing (also known as 3-D printing), and even unexpected changes in the global distribution of energy supplies, have implications for security and the economy in the United States as well as among many other trade-oriented nations, who may face a new and unexpected form of deglobalization.

This ability to produce customized goods locally, cheaply, and in smaller quantities may perhaps lead to a gradual loss of cost-of-labor advantages. Today, two-thirds of all Bangladeshi women working in the industrial sector—and one in seven young women from across the country—have jobs in the garment business.[4] It now provides 84 percent of the country's exports.[5] Three and a half million Vietnamese work in clothing production.[6] Localized advanced manufacturing could replace this traditional route to industrialization and economic development. Robots have been around for years, but robotics on a grand scale is just getting started: China today is the world's biggest buyer of robots but has only 97 per 10,000 workers; South Korea has 710.[7]

These advances also diffuse military power. Ubiquitous sensors, inexpensive and autonomous drones, nanoexplosives, and cheaper access to outer space through microsatellites all empower smaller states and even individuals, closing the gap between incumbent powers like the United States and prospective challengers. You can imagine a similar paradigm: more damages from less raw military strength. The proliferation of low-cost, high-performance weaponry enabled by advances in off-the-shelf software, navigation, and additive manufacturing diminishes the once-paramount powers of conventional military assets like aircraft carriers and fighter jets. This is a new global challenge, and it threatens to undermine US global military dominance, unless we can harness the new technologies to serve our own purposes. As we conduct ourselves throughout the world, we need to be cognizant that our words and deeds are not revealed to be backed by empty threats. At the same time, we face the challenge of proliferation of nuclear weapons, a still-existential threat that society has seemingly forgotten about.

Finally, the information and communications revolution is making governance everywhere more difficult. An analogue is the introduction of the printing press: as the price of that technology declined by 99 percent, the volume grew exponentially. But that process took ten times longer in the fifteenth, sixteenth, and seventeenth centuries than we see today with digital media.[8] Information is everywhere—some accurate, some deliberately inaccurate, such that entire categories of news or intelligence appear less trustworthy. The "population" of Facebook is nearly twice the population of the largest nation-state. We have ceaseless and instantaneous communication to virtually everybody, anywhere, at any time. Such access can be used to enlighten, and it can also be used to distort, intimidate, divide, and oppress.

On the one hand, autocrats increasingly are empowered by this electronic revolution, enabled to manipulate technologies to solidify their rule in ways far beyond their fondest dreams in times past. Yet individuals can now reach others with similar concerns around the earth. There is a sort of emergent order to this newly liquid marketplace of information and communication. People can easily discover what is going on, organize around it, and take collective action.

At present, many countries seek to govern over this diversity by attempting to suppress it, which exacerbates the problem by reducing trust in institutions. Elsewhere we see governments unable to lead, trapped in short-term reactions to the vocal interests that most effectively capture democratic infrastructures, even if they do not accurately represent the interests of most citizens. Both approaches are untenable. The problem of governing over diversity has taken on new dimensions.

The good news is that the United States is remarkably well positioned to ride this wave of change if we are careful and deliberate about it. Meanwhile, other countries will face these common challenges in their own way, shaped by their own capabilities and vulnerabilities. Many of the world's strongest nations today—our allies and otherwise—will struggle more than we will. The more we can understand other countries' situations, the stronger our foundation for constructive international engagement.

We have been studying these changes with a group here at Stanford University's Hoover Institution for the last few years now. We convened a series of papers and meetings examining how these technological, demographic, and societal changes are affecting countries and regions around the world, including Russia, China, Europe, Latin America, the Middle East, and Africa. After this, we turned our lens inward: what of the impact of these transformations on the United States—our democracy, our economy, our health, and our national security? For each issue, a broad swath of experts and practitioners from around the country and world have thoughtfully considered these emerging changes and traveled to Hoover to present their ideas through intense roundtable discussions over many months—but also to share them through community panels with the public, whose questions and reactions to these ideas we have found are often not the same as our own.

The more we study these revolutionary changes, the more interesting and important they seem to become, and the more clearly we see both the benefits they promise and the challenges they pose. As our friend Sam Nunn, the former senator from Georgia, has said, we've got to have a "balance" between optimism about what we can do with technology and realism about the dark side. So, we've aimed to understand these changes and develop strategies that both address the challenges and take advantage of the opportunities afforded by these transformations.

So, bound now at home by the COVID-19 crisis, we consider these forces, very much still in play, and look ahead. What have we learned about what awaits us on the other side of this hinge of history?

First is more migration—from the young and growing developing world to the aging and in many cases shrinking developed countries. Migration has many benefits. It allows the developed countries to maintain and expand workforces that would otherwise contract, to sustain their economies, and to support growing elderly populations. And migration provides productive opportunities for young people in countries with few jobs and those adversely affected by climate change. But migration must be prepared for and managed, with respect for the rule of law and an attempt to match the numbers and skills of migrants with the needs and capacities of receiving states.

Next is further automation and advancing technology throughout the economy, a phenomenon about which much has been said and written. Like migration, advanced technologies have many benefits, including increased productivity, a stronger economy, and a better standard of living. Most jobs will be redesigned, with some tasks assigned to machines. Machines can substitute for human work—but they can also complement it. Workers will need new skills; community colleges partnered with employers can play a major role, but reversing the shortfall in today's K–12 education is essential to providing everyone with a solid foundation to start. Our future national security also depends on taking advantage of revolutionary technologies while meeting the challenges posed by such technologies in the hands of adversaries.

A related phenomenon worth mentioning on its own is the growing use of additive manufacturing and other advanced manufacturing techniques to make things near where they will be used, and to provide flexibility in global trade patterns by introducing new degrees of freedom. In particular, this opportunity means the optimization of labor costs becomes less central to running a business more efficiently. Depending on policy priorities and where these capabilities are used, benefits might include lower production costs, the repatriation of manufacturing capabilities, more-secure suppliers, a reduced environmental footprint, and customized, even unique, products.

The ongoing information revolution, including social media and internet commerce, again can be expected to bring many benefits. But social media can also be used to spread disinformation, which is particularly problematic for democratic elections. We need to address privacy issues and objectionable content at home, and meet the challenge posed by adversaries that use technology to conduct information operations against us and to control their own citizens.

Meanwhile, the very environment around us is changing, undermining the stability and trend of general improvement that our country and others had generally taken for granted over the past century. So this is something new for us. Climate change is already having a real, observable impact on the environment. Infectious diseases are moving north, ecosystems are being damaged, and the incidence of extreme weather events is increasing. Governments should consider the costs they are likely to incur as a result of a changing climate, and what resilience investments might be effective to reduce them. Impacts may be more consequential in poorer parts of the world, due both to the weather changes these countries will experience and to the fewer resources or institutions available to smooth the process of adaptation and adjustment. The United States will need to undertake new efforts to reduce its carbon emissions. Given the huge scale of changes in investment and consumer behavior that will be necessary to significantly reduce emissions, policies to encourage that will have to be affordable at scale—for example, a revenue-neutral carbon tax and new investment in scientific research and technological innovation to reduce the price tag of clean energy and carbon-removal technologies.

As we look around the world today, we are convinced that the initiative to create what comes next has to once again come from the United States.

Why the United States?

The global changes explored throughout this book all have one thing in common. They are starving for good governance, at a time when we don't seem to have it. In fact, good governance is under attack.

Return to the bench we had to work with in building and mandating the postwar global commons. In a personal conversation with Deng Xiaoping in Beijing in the 1980s, Deng in a very straightforward way laid out China's plans on "opening"—both domestically, to allow people more freedom to move about the country and engage in commerce, and subsequently globally, through a greater opening to the outside world. "And I'm glad, Mr. Secretary, that there exists a reasonably coherent world to be opening up to," he said. It was a testament to the strength of leadership not just in the United States, but throughout Europe and other parts of Asia as well: Margaret Thatcher, Helmut Kohl, Bob Hawke, Yasuhiro Nakasone, Miguel de la Madrid, François Mitterrand, Brian Mulroney, Lee Kuan Yew.

Looking around the world today, we cannot say that we have that same depth of leadership. There is an angry sea. Britain is struggling with Brexit, and Europe seems to be tearing itself apart. The Middle East is more fraught—and more highly armed—than it was even during its turbulent years when Ronald Reagan was in office. Russia remains deeply troubled—a country with incredibly talented people, and thousands of nuclear weapons, but an economy smaller than Italy's. Some may point to a growing China as a potential stabilizing force, and it exhibits clear strengths. But in our own limited experiences it's hard to point to a time when China worked to develop the international order in a way that was not guided toward the direct self-interest of its ruling party. Even domestically, the interests of the party eclipse those of its own people, as exhibited by its handling of the 2019 protests in Hong Kong. And as we look on the other side of this hinge of history, with changing technologies and movements of people, it's not clear that the interests of the Chinese Communist Party would have any more universal appeal than they have in the past, with willing cosignatories. Democracies, among their many benefits, are not hobbled in this way. On the international stage, for all of China's interest in establishing its own realm of influence in the Pacific and beyond, it has not had much success in making friends, even with the neighbors.

Everywhere we look, governments face the challenge of capturing the benefits of changing demographics and advancing technology while overcoming the problems. And the United States is best positioned to take advantage of these global transformations; we have long experience in diversity, and our companies and universities lead the commercial development of twenty-first-century technologies. We can face the future with confidence, provided we take the necessary steps: prepare a process for welcoming migrants in ways that help them assimilate as productive Americans while preserving the concept of citizenship; reverse the decline of K–12 education, which otherwise sharply limits our own children; adapt our social safety net to meet changing needs; protect our democratic processes; address vulnerabilities to information and cyber threats; and exploit the military and broader security applications of technology to our net benefit. As diversity increases, good governance becomes both more difficult and more necessary. We must find the right combination of traditional, market-based policies that have proven to work to build a prosperous society, along with practical solutions to problems without precedent. In many cases the individual states can take the lead, and serve as laboratories to develop innovative approaches.

Meanwhile, the developed countries, with leadership from the United States, can cooperate to prepare for migration, and address problems that do not respect borders, such as privacy, disinformation, and threats to democracy. They can support developing countries, with an emphasis on supporting women, in areas such as education (particularly for girls), economic opportunity, and health and child care, all of which help reduce fertility and support economic growth.

This is a likely a sentiment with which the late Senator John McCain would have agreed. A couple years ago, Henry Kissinger and your authors traveled to Washington for testimony to the Senate Armed Services Committee, which McCain chaired. Unfortunately his illness kept him watching from home in Arizona instead. In Dr. Kissinger's remarks, though, he praised the chairman not just as a defender of American values but also as "someone who, whenever the weak were threatened, made it clear that America was on their side." Our changing world is not necessarily one of a nature that will pit strong nations against the weak. Instead we will see in it a host of novel and shared challenges: Complex technologies whose implications are hard to anticipate, or even measure, and that will move fast. New environmental risks that may unfold in unforeseen ways. A mass movement of people around the globe, in some instances to countries with little modern experience in accepting migrants of any stripe. America has the size, the prosperity, the institutions, and, we believe, the will of its people to lead both by example and through direct cooperation with those nations who will struggle the most with this governance gap.

One early 1980s snowed-in weekend supper at the White House, hours after one of your authors had landed from a diplomatic trip to China, marked the first time that President Reagan made anyone aware of his desire to do something constructive with the Soviets. The Cold War was as cold as it could get, and every expert and every agency had its own concept of how things should proceed in that relationship. There was "détente." There was "linkage." None of these were satisfying. There was great tension, everything was frozen.

But President Reagan understood the direction he wanted to go, and that governance meant doing what made sense at the time to get there. He didn't need the perfect grand strategy or policy concept to do that. Over an impromptu dinner, in a city otherwise closed by a chance winter storm, through his questions about the Chinese, one could sense his interest in getting a personal feel for how communist leaders thought and acted. He wanted to meet with the Soviets. As a secretary of state, to have a chance meeting like this that showed which way the president wanted to go was a giant revelation. It was possible to see that he and the First Lady had talked this over, and it was what they truly thought. The Defense Department and the CIA didn't buy it, so they were opposed. But understanding a leader's personal convictions gave the freedom then to begin to work in earnest with the Soviets on things that both countries felt to be worthwhile. Others around the world, meanwhile, saw what we were doing and saw how they could be a part of it. And over time that momentum helped lead to the end of the Cold War.

America can help the world through this new hinge of history through leadership—recognizing a principle, unglamorously figuring out how to get started on a path that makes sense, and bringing people on board as you go. Today too, one could expect great institutional opposition in government, across countries—whether allies or adversaries—to picking up on these changes and doing something about it. So you want to be ready when things do open up, just like that snowed-in weekend in the early 1980s. In our experience it takes a jolt to get people to act. It's early in our current pandemic crisis, but maybe this is that moment. When it comes, if you are ready, you get something done. If you are not ready, then the situation goes by, and nothing happens.

The changes we can now see across the world—looking at the shape of its populations, its technologies, or the climate—aren't just affecting one person. They affect everybody. Everyone has a stake, so let's see if we can't do something about them that is constructive. This has happened before over the past century, when people felt they had an opportunity to change things—they did, and it helped. As Americans, we can once again take another careful look at the world around us, and the world ahead, so that we can find constructive things to set out doing in it.

PART

# II

# A Walk around the Emerging New World

We will come back to how America will fare on the other side of the hinge of history. But let's begin by first taking a walk around this emerging new world, where the experience of the United States is just one piece of the puzzle.

As we surveyed this subject over the past few years, it's fair to say that every area expert or subject practitioner who stopped through the Stanford campus in Palo Alto, no matter the depth of their experience, told us that they were struck by one recurring theme. And that theme was demographics. The composition of global populations, and the economic or governance stories that those peoples create, are the foundations on which all other global dramas play out. That demographic picture of the world is changing rapidly and greatly, with truly significant consequences—and, for the most part, we have found that change to be ignored or simply unknown.

Here at Hoover, we were able to learn from an economic demographer and author among our ranks named Adele Hayutin, who uses charts and other visualizations to teach investors and policy makers about the shape of societies around them. Her explanations animated each of the regions we contemplated, and which we now will examine with you in the following pages.[9]

In short, essentially all wealthy countries have for decades had below-replacement rates of fertility and rising longevity. Most are therefore losing working-age populations, some rapidly, resulting in the dramatic transformations we see in Figure 1. The United States, Canada, and Australia are exceptions: they continue to have rising working-age populations, but only because they are countries with high immigration. In the United States, we are lucky to have a wide variety of immigrants from many countries around the world and many backgrounds. Some of our immigrants pick strawberries, valuable and

**FIGURE 1 Global population shift to Africa will accelerate.** Fueled by high fertility, Africa is projected to account for nearly all the projected global population growth from 2020 to century's end. Its share of global population will increase from 17 percent in 2020 to 26 percent in 2050, and 39 percent by century's end. Asia's share falls from 60 percent in 2020 to 43 percent by 2100.

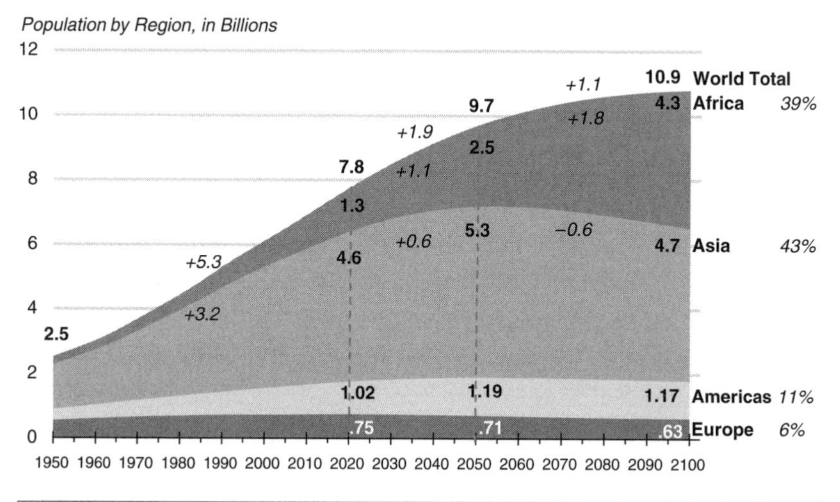


Data Source: United Nations, *World Population Prospects: The 2019 Revision*, medium variant. Demographic analysis by Adele Hayutin, 2020.

necessary work in many parts of this country. Some win Nobel Prizes: more than one-third of prizes awarded to Americans over the past hundred years, and nearly two-fifths in the past twenty, have gone to immigrants.

Meanwhile, poor countries with high fertility rates are all too often places where governance is weak, where economic prospects are poor, and where there is a large dependence on agriculture or other primary economic activity. Where these countries lack the capacity to adapt, they are vulnerable to chaotic security situations (potentially fueled by new forms of technology-enabled off-the-shelf weaponry) as well as to the ravages of climate change (including droughts that produce famine). In other words, migration will have to occur. Some will be internal, some will not. Where will these migrants go?

Of course each country must manage the impacts of its own demographic patterns—the fiscal implications of aging societies, for one. But those demographic patterns and choices also now have broader international scope. For example, geographic redistribution of the world's workers, and what levels of education and productivity they bring to a high-technology manufacturing environment, will affect trade and geo-economic relationships. And as for international migration, what is the role of migrants in adding to, versus subtracting from, the prosperity of a receiving country?

These demographic developments put us on a hinge of history. Let's compare how this situation differs from what we saw at the end of World War II, and where we are heading next.

## Then, and Now

As Figure 1 shows, in 1950 the global population totaled 2.5 billion, less than one-third of today's population. Asia, with 1.4 billion people, accounted for more than half the total, while Europe accounted for 22 percent. Africa's population in 1950, 228 million, was just a sliver of the world's total, accounting for only 9 percent.

That world has been described as comprising two major demographic groups—wealthy industrialized countries with low fertility and high life expectancy, and poor, mostly agrarian countries with high fertility and relatively low life expectancy. About one-third of the global population lived in those wealthier regions, primarily Europe and North America. And two-thirds lived in Asia, Africa, and Latin America—the poor countries in those three regions often described as Third World countries. They had high fertility rates, hovering around six births per woman, and relatively low life expectancy, averaging just forty-two years.

The demographic transformation since then has been dramatic. Global population has more than tripled, reaching 7.8 billion. The population has continued to shift toward the developing countries. Africa's population has increased nearly sixfold, and its share of world population has increased from 9 to 17 percent. Asia's population has more than tripled, and its global share has increased to 60 percent. In contrast, Europe's population has grown by less than half, reaching 748 million in 2020, with its global share decreasing to 10 percent. The global fertility rate peaked around 1965, at five births per woman, and has since fallen by half. At the same time, worldwide life expectancy has increased from forty-seven to seventy-two years, with an average gain of sixteen years in high-income countries and twenty-eight years in low-income countries. Although fertility has declined almost everywhere and life expectancy has risen in most places, the pace and timing of these shifts have differed, leading to divergent demographic outcomes: varied age structures, uneven population growth, and, more important, uneven economic growth.

The 1950 dichotomy of the low- and high-fertility groups has given way to much more demographic diversity, determined largely by the pace and timing of the fertility and life expectancy changes from one country to the next. In 2020, there is a rich demographic variety across countries that rewards examining each in turn; the chapters that follow do just that. To put those in context, however, we can start by generalizing across three demographic groups that have emerged through the broad changes since the 1950s:

- First are the high-income countries that have had a long history of below-replacement fertility and now face shrinking workforces and aging populations.
- A second group includes countries that have had large fertility declines since the 1960s and now have fertility rates near replacement rate. Many countries in this group had extremely steep fertility declines that are already showing up in shrinking workforces. In other countries the fertility declines were more gradual.
- A third group includes the least developed countries, mostly in Africa, where fertility rates, although declining, are still very high and where population growth remains explosive.

Let's take those in turn. Today's wealthy countries have experienced below-replacement-rate fertility for four decades. The rate averaged around three births per woman during the 1950s and 1960s but had fallen to 2.1 by 1980. Fewer children were born, and in turn, growth in the working-age population slowed. At the same time, these countries also had slow but continual gains in life expectancy, so both the number and share of older people dramatically increased. Over the second half of the twentieth century, the ratio of working-age population to dependent-age population (that is, both the very old and the very young) increased, and that increase provided a boost to economic growth. That ratio has since been declining, which makes it more challenging for governments to provide desired social services, including pensions and health care, and also afford other productive investments. Most high-income countries, after decades of low fertility, will soon face shrinking workforces and declining total populations. But there are important exceptions—countries where immigration has boosted workforce and total population growth, including the United States, Canada, and Australia. (Immigration has also occurred in Russia and Germany, but not enough to offset the natural population decline.)

Second are those previously "developing," now middle-income, countries, which fall into two subgroups: those with early and rapid fertility declines, and those with more gradual declines. The combination of falling fertility rates and increased life expectancy in many developing countries has led to slower population growth, slower workforce growth, and rapid population aging. Countries in this group that had the earliest and most precipitous fertility declines will see declining workforces in the next decade, while others that had more gradual fertility declines will see continued growth for a decade or two, followed by a drop-off in workforce growth. China and South Korea, for example, had fertility declines that were among the steepest and earliest globally (partially self-inflicted by bad policy in the Chinese case), and these two countries are projected to see declines in their working-age populations over the next twenty years, with no long-term reversal in sight. Chinese fertility fell from a rate of over six births per women in the late 1960s to less than half that within a single decade, and it halved again over the following two decades. Iran's fertility decline, though even steeper than China's, occurred later, so its workforce is not projected to start contracting until around 2040.

The third demographic group, the poor countries, has the highest fertility rates, averaging 4.5 births per woman. For these countries, population growth is strapped to a rocket. Growth in working-age population continues to outpace economic prospects, especially in areas with chronically subpar governance. These populations also face challenges of pollution, climate change, and natural disasters. Even more important, growth in this group has been disproportionately fast: today's low-income countries account for 10 percent of the world population, but are projected to account for 36 percent of population growth over the next thirty years. Most of these low-income countries are further categorized by the United Nations (UN) as "least developed," meaning they have the least capacity to respond to economic and environmental challenges.

Many of the demographic changes we anticipate over the coming years, including continued population growth, rapid aging, shrinking workforces, increasing urbanization, and accelerating international migration, stem from population trends set in motion since 1950.[10] Indeed, many of the current challenges stem from two major demographic successes: the large drop in global average fertility, which has dampened explosive population growth in many regions, and the remarkable increase in life expectancy, which means that more people are surviving childhood and living to older ages.

## Rapid Growth in Africa

One consequence of the divergent fertility and longevity trends, as should be clear from Figure 1, is the shift in composition of the world's population toward Africa. Not only did Africa have the highest fertility rate of all the regions—6.7 births per woman—back in 1965, but the fertility rate stayed high, above 6 births per woman, until the early 1990s, falling only gradually since then. The regional average in 2020 is 4.4, resulting in radical population growth. Nigeria is a particularly notable example, its explosive growth setting it on a course to soon surpass the United States in population: its 2020 count of 206 million people is projected to double over the next thirty years. In contrast, the fertility declines in Asia and Latin America started earlier, were larger, and occurred much more quickly. Asia's fertility rate is now 2.2 births per woman, down from 5.8 in 1965, while fertility in Latin America has fallen below replacement to 2.0, from a peak of 5.9.

Over the past seventy years Africa's population has increased almost sixfold and Asia's population has more than tripled. By 2020 we see that Africa's share of global population has nearly doubled, from 9 to 17 percent, while Asia's share has increased from 55 to 60 percent. In contrast, Europe's population has grown slowly, and its share has declined, from 22 percent in 1950 to just 10 percent today.

This geographic shift is projected to continue. Due to its still-high fertility rate and continued explosive population growth, Africa's share of the global population is projected to increase from 17 percent in 2020 to 26 percent by midcentury and 39 percent by the end of the century. In contrast to Africa's continued rapid growth, Asia has seen its population growth dampened by significant fertility declines and is projected to peak at 5.29 billion people around midcentury, then decline by 571 million over the second half of the century. Consequently, Asia's share of global population will fall from 60 percent in 2020 to 54 percent in 2050, and finally to just 43 percent by 2100. Most of Asia's population decline after 2050 will occur in China.

More than 90 percent of the world's total population growth since 1950 has occurred in developing regions, which now represent 84 percent of the global population. And going forward, almost all the projected global population growth will occur in the *least* developed of those developing countries—those with the least economic capacity and weakest social infrastructure. According to the medium variant United Nations projection (which extrapolates moderate continued fertility declines), the total population of those forty-seven least developed countries will nearly double, from 1.1 billion in 2020 to 1.9 billion, by 2050 (or 19 percent of the world's population), and grow to 3.0 billion by the end of the century.

It's worth noting that almost all of that growth will occur in urban or urbanizing areas. Urban population will grow the fastest in the poorest countries, and because most urban growth is concentrated in coastal areas, these growing populations are vulnerable to environmental threats. Unmanaged growth will exacerbate stress on urban infrastructures, and without sufficient jobs, housing, and security this could increase the movement of people in search of safe locations and better economic prospects.

In short, demographic trends across this new hinge of history set out Africa as a huge new global frontier. But a lot depends on governance. Will the Nigeria that adds more than half a billion citizens by the end of the century look more like Indonesia, with stability and a reasonable base for social and economic opportunities, or will it look more like Zimbabwe does today, a failed state with hyperinflation and a legacy of productivity-sapping corruption? If Africa's economy were to start growing like India's, it would be on a path to overall economic parity with the United States by midcentury, and would earn the global influence that goes along with that. Or, if its exploding youth population were to start producing emigrants to Europe in equal share to how a growing but troubled Mexico did to the United States in the 1990s, then that continent could be facing not just millions of refugees, but hundreds of millions of economic immigrants. Security in the African continent long took backstage to the power rivalries or other global crises that played out in the postwar period in Asia and Europe: as Africa grows to assume a simply larger share of everything that happens in this emerging new world, that will change. The United States should see this coming, and whatever choices it makes to guide that process, it should know that it will be affected by the outcome.

## Changing Age Structures: Graying Populations and Shrinking Workforces

Not only are populations shifting around the world, but the character of those populations are different now, too. The more-developed countries have had below-replacement fertility rates for several decades, and at the same time they have enjoyed increased longevity. Their populations are aging, and their population growth has slowed or even reversed. Most importantly, as Figure 2 illustrates, the decline in the number of children has resulted in much slower growth in working-age populations, and many countries are already subject to workforce contraction. Over the next forty years, eight out of the fifteen largest economies will see shrinking workforces. Just over the next twenty years, Japan will witness a 19 percent decline in its working-age population; Germany, 13 percent; and Russia, 8 percent. China, the globe's second-largest economy and previously the factory floor to the world, is projected to have an 11 percent decline in working-age population over the next twenty years, and its total population is projected to begin declining by 2030.

The contrast between China and India shows the impact of differences in the two countries' underlying trends. China, which had a steep fertility decline beginning in 1965, has already peaked in working-age population; its pool of workers is projected to decrease by 174 million by 2050. In India, where the fertility decline was gradual and just recently fell to replacement rate, the working-age population will continue to grow, adding 183 million over the next thirty years.

The United States is an exception to the overall developed-country trend—our population is fueled by immigration, so it is still growing, and the continued increase in the number of children is in turn fueling workforce growth. Though the growth rate is slower than in the past, the working-age population is projected to increase by 4 percent across the next twenty years, and another 4 percent across the subsequent twenty. Among the advanced economies, the only countries that have growing workforces are those that benefit from immigration, including the United States, Canada, United Kingdom, and Australia.

At the same time as workforce growth is slowing, and in many cases shrinking, the advanced economies face the increasing fiscal and social challenges of graying populations. For the more developed countries overall, the proportion of the population aged sixty-five and older is projected to increase from 19 percent today to 27 percent by midcentury. In Japan, the proportion of the population aged sixty-five and older is projected to swell from 28 percent in 2020 to 38 percent by midcentury. South Korea has had a fertility rate well below replacement for several decades and is now the fastest-aging country in the world. It is projected to see a 23 percent decline in its working-age population over the next twenty years, and it will see a

**FIGURE 2 Eight of the fifteen largest economies face shrinking workforces.** The United States, Canada, and Australia are exceptions, largely due to immigration. Worldwide workforce growth slows by half; due to large fertility declines, growth in Mexico and India will turn negative after 2050.

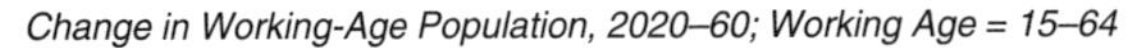

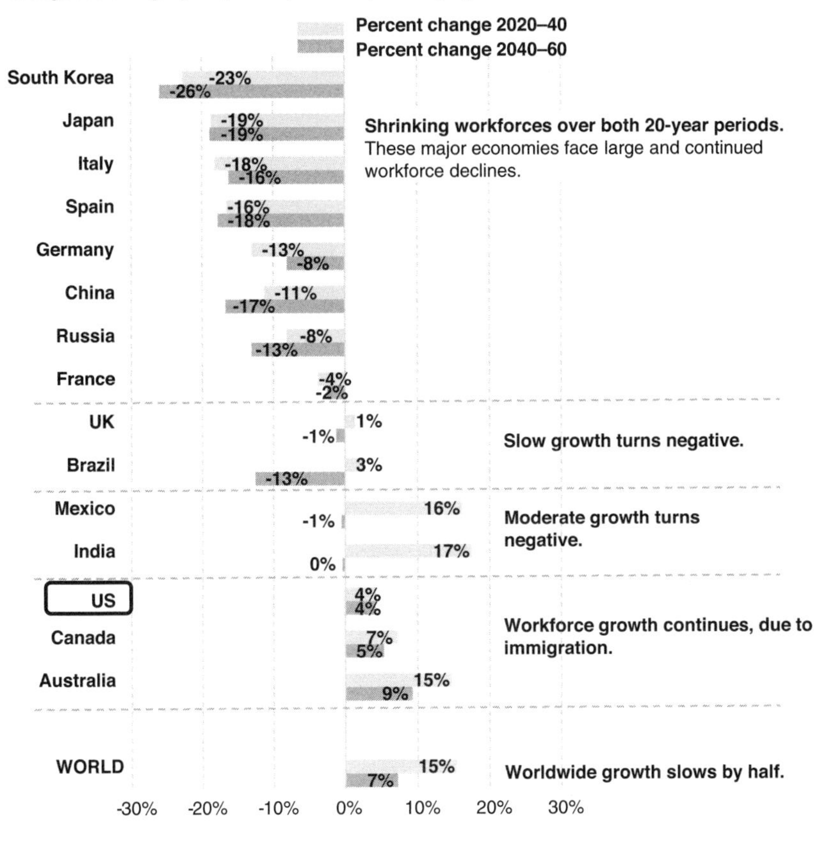


Data Source: United Nations, *World Population Prospects: The 2019 Revision*, medium variant. Demographic analysis by Adele Hayutin, 2020.

notable rise in its proportion of older people, from 16 percent in 2020 to 38 percent in 2050.

Again, the United States is an exception—as Figure 3 shows, we are comparatively young. We are rightly concerned about aging baby boomers and retirement costs, but our aging population (17 percent of the total as of this writing) is much smaller and growing more slowly than in other countries. The share will rise to just 22 percent by midcentury. Compare this to China,

**FIGURE 3 The pace of aging will accelerate over the next thirty years.** South Korea, China, and Iran have especially steep increases in their older populations. In contrast, the United States is aging more gradually and will remain relatively young among major economies.

*Percentage of Population Aged 65+, Selected Countries*

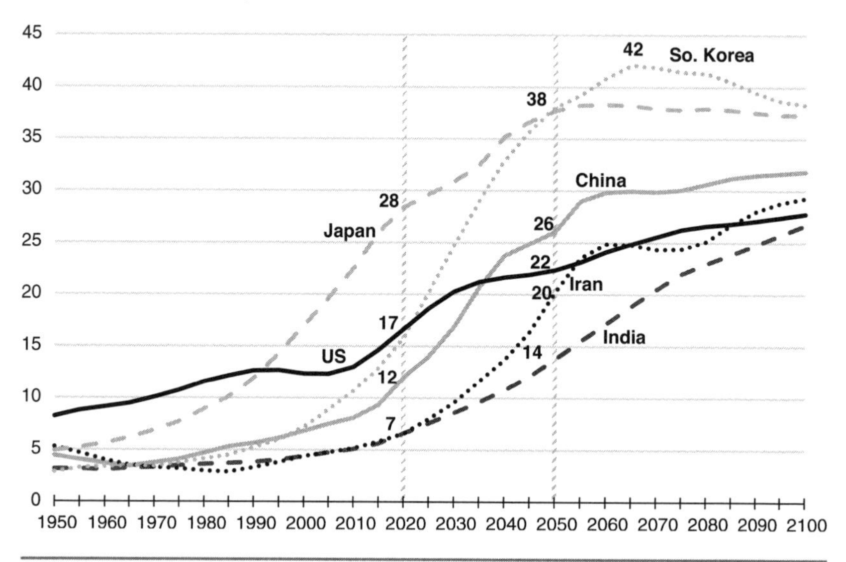


Data Source: United Nations, *World Population Prospects: The 2019 Revision,* medium variant. Demographic analysis by Adele Hayutin, 2020.

where by midcentury 26 percent of the population will be aged sixty-five and older, up from just 12 percent in 2020.

Iran is another example of a country with a very rapid fertility decline—similar in steepness to China's—but it began around 1985, about twenty years later than in China. Iran is rapidly aging. As recently as 1990, Iran had a median age of seventeen years, typical of a young and growing population; but by 2020, its median age has increased to thirty-two. By 2050, the median age will be forty, making the population of Iran just a few years younger than that of the United States, which by then will have a median age of just forty-three. Iran will also have a huge increase in the proportion of older people, swelling from 7 percent in 2020 to 20 percent in 2050.

Contracting workforces and graying populations pose huge challenges for social welfare. These two trends suggest that work will *need* to be done differently in the future to maintain economic well-being. At the same time, technological advances are changing the way work *can* be done, so we need to pay attention to the intersection of the demography of slower workforce growth and the potential of technology to improve productivity and the general interest.

As we discuss elsewhere in this book, a strong basic educational foundation is the key factor for societies looking to take advantage of such technologies—to get those needed productivity gains, boost workforce-participation rates, and mitigate the shrinking labor pools that we know to be coming.

Meanwhile, these graying countries, especially those that do have high education levels, can make reforms now to dampen the impact of declining workforces: incentivizing extended retirement ages for willing and productive workers, and encouraging women of all ages to join or remain in the workforce.

A key point is that the United States is in better shape demographically than other large economies, and if we pay attention, we can capitalize on our demographic strengths.

## Increased Migration

Poor economic prospects in rapidly growing countries, political instability in many of these same areas, and vulnerability to climate-driven threats—combined with ubiquitous access to long-distance communication and information about better conditions elsewhere, and even the open desire of aging wealthy countries to attract young workforces—will spur increased migration both internally and across international boundaries.

The number of people living outside their country of origin in 2019 increased to 272 million, or 3.5 percent of the world's total population.[11] This is up from 250 million in 2015. For some individual countries, the impact of migration on population growth and composition can be significant and the implications for social and political stability enormous.

The United States is already by far the largest destination country, hosting fifty-one million foreign-born people, followed by Germany, Saudi Arabia, and Russia. The top ten host countries also include the United Kingdom, Canada, and Australia—three countries where immigration contributes to growth in the working-age population. Globally, the top sending countries in 2019 were India, Mexico, China, Russia, Syria, and Bangladesh. (Notably, Russia is both a top sending and a top receiving country.)

Political refugees have also been on the rise—including both those who cross international borders and internally displaced people who remain within their country of origin. According to the UN, the number of refugees reached 25.9 million worldwide at the end of 2018, the highest number on record. The top five countries of origin included Syria, Afghanistan, South Sudan, Myanmar, and Somalia. In addition, an estimated 41.3 million people were internally displaced by conflict and violence. That number has more than doubled since 2000, with a sharp increase since the Syrian Civil War began in 2011.

These upticks in migration, whatever the source, have significant economic and political implications that stretch well beyond the countries of origin or destination. In addition to economic and humanitarian concerns, global migration, particularly when episodic or unmanaged, has significant implications for global security and citizenship. With these trends likely to continue, the situation cries out for a thoughtful governance response.

Receiving countries should take proactive steps to target desired year-to-year immigration-flow levels—not too high in surge years, and not too low in periods of need—in order to complement domestic demographic and economic trends. Deliberating these values in advance will help prepare for the next border crisis. Going further, receiving countries could consider moving upstream—guiding the origination of potential migrants while they are still within their source countries. This could include measures both to limit the generation of economic migrants and to better prepare the ones who do leave to assimilate and productively contribute to receiving countries. Such efforts might include investments in K–12 education and training, anti-corruption, and security, and even in improving the capacity to absorb natural disasters, crop failures, or other weather impacts that otherwise produce migrant pulses.

The domestic problems of countries producing many emigrants are not the responsibility of receiving countries like the United States to solve. What's more, receiving countries do not have the resources to do so at scale even if they wished to. At the same time, receiving countries have a very real stake in what happens when their citizens may become our residents, bringing their own capabilities or needs with them. We are affected, whether we like it or not. For example, many of the migrants who come to the United States today are among the most highly educated in our society; yet many others lack even a high school degree, or English language ability. Uneducated migrants can perform needed manual labor once they arrive, but their prospects are otherwise limited; it falls to our own K–12 school systems to ensure that their children, as Americans, improve their productivity well beyond that. We think that there is a role for private, nonprofit, and multilateral development programs in addressing each step of this migrant generation chain at its source in poor countries abroad and that doing so would provide a clarity of purpose, and better US public support, in such programs' own development missions.

Let's work from this demographic foundation to continue the walk around the world region by region, learning from our project's collected experts and practitioners of government. First, an overview, and then more detail on each in the chapters that follow.

Start with **Russia**. An assessment of the strengths and weaknesses of the Russian Federation as it addresses the coming demographic, economic, and technological challenges can be a first step toward the development of a strategy to deal with it in the emerging new world. And Russia under President Putin looks weak. In our discussions, Stanford professor and longtime Russian interlocutor David Holloway explained how Putin has focused his governance efforts on preserving stability rather than modernizing. Although Putin speaks of the need to grow and adapt to this new world, he does not appear to be taking the necessary steps to do so. Meanwhile, as described by Soviet historian Stephen Kotkin, we see a continued turn toward greater authoritarian rule in Russia, with the Putin administration wielding new technologies in the service of its own military and political ambitions. But it remains to be seen whether an inherently brittle regime such as Russia's can overcome the looming technological and social challenges described above. Michael McFaul, former US ambassador to Moscow, built on these perspectives to argue that Putin's admiration of Chinese autocratic power has led him to mistake that as being the foundation for that country's strong economic performance—and Putin's attempts to emulate this at home are hobbling Russia's technological capabilities and the economic potential of its citizens. Meanwhile, a group of prominent Russian politicians and academics, including the former foreign minister Igor Ivanov, helped us also consider how Russia sees its own role in a changing world, arguing for a new international system of governance to address the migration of people and technological revolutions. Anatoly Vishnevsky, of Russia's National Research University

Higher School of Economics, explained how we are witnessing hemispheric demographic trends in that country—an aging one to the north and a rapidly growing one to the south. And Moscow-based Primakov National Research Institute of World Economy and International Relations scholar Ivan Danilin questioned whether Russia can keep up with the military technological superiority of the United States and China. The next chapter explores each of these concepts more deeply.

Next is **China**, which has higher global ambitions given its capabilities, but which also faces serious domestic challenges that may well occupy its attention. Our analysis of China explored that country's poor demographic outlook, the use of social media and communications technology by its people and government, and its development of advanced technologies for both economic and military gain. On the back of Deng Xiaoping's economic openings and the ingenuity and hard work of its exceptional people, China has enjoyed a period of extraordinary growth and is becoming a world leader in some key twenty-first-century technologies. But it must now contend with an aging society and shrinking workforce, robbing it of the fuel that propelled years of productivity gains as young, rural migrants became skilled, urban workers. Moreover, China's authoritarian leadership wrestles with the governance choices posed by the economic imperative for further opening and with dynamism versus reflexive statist crackdowns on firms, on local leaders, and on social freedoms as new stresses emerge. Take the traditional *hukou* system, which classifies Chinese citizens into regional affiliations and is necessary for their receiving essential government services such as education or formal employment. Dramatically loosening these classifiers would help unleash the sort of dynamism that US citizens enjoy, to be able to move across states to new jobs, to new educational or personal investment opportunities, or to escape poor conditions. But doing so would mean giving up a vestige of state control over people's lives that successive Chinese leaders have been shy to let go.

Technology plays directly into this mix. In the year 2000, Chairman Jiang Zemin's contribution to modern Chinese communist ideology was the concept of "three represents": that is, as nonstate forces in Chinese society such as private enterprise grew stronger, the Chinese Communist Party, in order to stay relevant, would have to open its doors wider to welcome the participation of nontraditional members like entrepreneurs and managers. At that time, the Chinese economy was seemingly growing faster and people's day-to-day lives were more complex and varied than overstretched party cadres could ever hope to control. But technology changed all that in just a few years. Today Xi Jinping's Chinese state automatically surveils and collates the personal utterances, friend networks, purchases, ideology, and travel choices of every Chinese citizen in ways that chairmen Deng or Jiang would have never considered. As a result, the Party no longer worries about whether or not businesspeople will join its ranks—instead it simply forces even private firms to accept Party members to sit on their boards. From communications to manufacturing and infrastructure, how will China's pursuit of next-generation technologies be used by its government to try to extend its otherwise slowing domestic economic development and establish more enduring global partnerships?

As part of our project, Nicholas Eberstadt, of the American Enterprise Institute, examined China's demographic outlook, identifying not only a coming labor-force contraction and a graying society but also long-term problems arising from urbanization, internal migration, and shifting family structures. And he found that demographic constraints appear poised to hinder both China's economic growth and its geopolitical aspirations. Military technologist Elsa Kania meanwhile considered whether there is an "AI arms race" between China and the United States. As both countries develop military applications of AI and other new technologies, such as additive manufacturing, we expect a growing military challenge; in the near term, the uncertainty and limitations of AI likely pose a risk. Looking to the civilian side of technological development, China-based technologist and venture capitalist Kai-Fu Lee described how artificial intelligence is transitioning from an "age of discovery" to an "age of implementation." In this next phase of development, which he argued will favor Chinese capabilities, companies will apply AI to an expanding set of tasks, but its widespread application threatens to exacerbate inequality and favor monopolies. And what of China's formidable information state? Maria Repnikova, assistant professor at Georgia State University, reviewed with us how the Chinese and their government have shaped social media use to their purposes. She argued that while the government has clearly employed these tools to further authoritarian ends, it has also used them to become more responsive to an increasingly connected and active citizenry. Finally, former US ambassador to China Stapleton Roy detailed how the Chinese government plans to become a global leader in these emerging technologies and what that growing economic, military, and technological capacity might mean for the United States.

China faces a number of domestic obstacles, especially in its demographic outlook, and its ruling Communist Party moreover lacks the confidence to abide the continued success of Chinese people's self-determination under other systems, as in Hong Kong and Taiwan. This is a troubling mix. In this complex interplay, there is nonetheless scope for constructive engagement between our two countries in areas of mutual benefit—without having to overlook substantial ongoing disagreements. Mature leadership, while continuing sober and quiet diplomacy in the background, points the way forward on both sides of the Pacific.

Foreign policy starts in the neighborhood. For us in the United States, this means viewing the changing **Latin American** landscape through the lens of underlying demographic and technological trends, and pondering how the United States may usefully engage in the region as it navigates similar forces of history. An old friend, the former Mexican finance minister Pedro Aspe, adeptly arranged a roundtable discussion on this question. On demographics, labor forces in Mexico, Central America, and South America continue to expand, in many countries at rates well above the global average. But changes are underway, with the current rate of workforce expansion in Mexico, for example, already half what it was twenty years ago, and trending toward zero growth by midcentury. What will be the domestic-labor, governance, and international-migration implications of this dramatic break with historical norms?

With Aspe's help, we were fortunate to host a group of demographic experts from El Colegio de México in Mexico City on a trip to Palo Alto to discuss these consequential but underappreciated demographic shifts. They observed how population dynamics point to Mexico maintaining its current net-migration levels near zero, but also to continued or even rising rates of youth migration from Central America through Mexico toward the United States, given poor conditions in the region. Responding to this change will mean new policy choices for our southern neighbor. This of course highlights the issue of Central American living conditions. Former Guatemalan minister of the economy and of public finance Richard Aitkenhead and technology entrepreneur Ben Sywulka put together an analysis looking at the specific conditions of Central American countries to ask what opportunities twenty-first-century technologies may offer in this region of generally weak governments, poor institutions, and underdeveloped labor markets. They argued that digital mobile platforms that efficiently connect providers of goods and services to potential customers offer the chance to leapfrog institutional development, broaden labor participation, and reduce the prevalence of informal sector jobs (those that operate outside of any regulation or tax regimes). It is an optimistic vision of how a region that is not at the forefront of technological advances may nonetheless be able to adapt these tools to local circumstances, with dramatic impact. We hope that creative efforts such as these to use new technologies to overcome the stubborn working and living conditions of the people of Central America, particularly the troubled Northern Triangle—comprising Guatemala, El Salvador, and Honduras—can be successful.

Moving south, Chilean politician Ernesto Silva devoted his time visiting the Hoover Institution to observing for us the historical roots of and current trends in governance of Latin America, particularly South American states. Silva decried the tendency for electorates to fall back on "strongman" caudillos (military dictators) over sustainable institutional reform and development, and he wonders how a growing, politically moderate, consuming middle class across many Latin American countries will intersect with the dramatic adoption of digital communications and social media, which offer a direct connection from politician to citizen. Can these tools be used to improve governance and lead to economic growth and better living conditions, or will they be limited to popular appeals meant to excite voters only during elections? This dynamic, so well observed by Silva, dramatically came to a head in his home country just a few months after his visits with us. Chile, once the poster child for economic and social success on the continent, found itself mired in wildcat protests and the throes of a new constitutional convention amid widespread self-doubt about its achievements. Here is a country that decades ago, in desperation to do something about its poor economy, adopted free markets, and it worked; though still under the rule of a strongman dictator at that time, citizens who had first tasted economic freedom also craved political freedom, and the surprising elections of 1990 brought real change in that direction. Now, technologies seem to be driving new demands, with troubling economic implications, and to an uncertain end.

Then there is the **Middle East**, where so much US attention has been focused since the end of the Cold War. To outside observers, the Middle East and North Africa have long been defined by violence, political upheaval, and religious conflict. Such a view is not unfounded; from the Iranian Revolution to the Arab Spring, the Iran-Iraq War to the Syrian Civil War, the region has seen more than its fair share of turmoil in recent decades. But it is inaccurate to reduce it to a battlefield of old feuds and new wars. States across the Middle East and North Africa confront many of the same global concerns as all other nations and regions. They must contend with the global challenges of shifting demographics, mass migration, governance in the internet age, and the impacts of greenhouse gases, which are affecting the region's already harsh climate. Yet they also face new opportunities, such as the economic potential of new technologies and the expanding role of women in their economies. We asked what these global transformations mean for Saudi Arabia, Turkey, Iran, Egypt, and Israel and considered the state of technological development and entrepreneurship in the region. Against the backdrop of the Sunni-Shia divide, the Israel-Palestine conflict, and other long-standing sources of political dispute, can Middle Eastern and North African states adapt to the challenges and opportunities posed by a rapidly changing world?

On the Arabian Peninsula, Harvard University scholar (and unwilling Moroccan prince) Hicham Alaoui sees a mismatch between Mohammed bin Salman's autocratic rule and Saudi Arabia's large, technologically capable youth population. Perhaps the kingdom can move toward a pluralistic society and allow its dynamic younger generation to flourish, but such a move could introduce political instability, a prospect abhorrent to the crown prince. In North Africa, Houssem Aoudi, a Tunisian entrepreneur and founder of a variety of local start-ups, reviewed the state of technological development and entrepreneurship in the United Arab Emirates, Egypt, and Tunisia, concluding that the region has yet to "climb from imitation to innovation." Stronger political and civil support, both within and across states, is necessary for entrepreneurs to make that jump. In turn, Egypt, argued Stanford University professor Lisa Blaydes, has demonstrated some good governance—for example, in aspects of public health—but it has struggled to generate job opportunities for its growing youth population or to address endemic social challenges such as rampant obesity. If these trends continue, Egypt may find itself in the midst of renewed political instability or upheaval. Israel, as conceptualized by Arye Carmon, founder of the Israel Democracy Institute, has built powerful "hardware"—durable institutions, a strong military, and a dynamic, high-tech economy—but it continues to lack good "software," namely a constitution. Israel is a diverse, at times disharmonious society; strengthening the political traditions and cohesion of the state will be key to the state's continued success. Former Turkish parliamentarian Aykan Erdemir, meanwhile, warned that Turkey is at a "crossroads": the current government could further centralize its economic and political system in a bid to cement its own power, and in the process waste the nation's already-narrowing window for economic growth fueled by favorable workforce demographics, or it could loosen control and encourage civic growth. Turkey's NATO allies, the United States in particular, would do well to encourage the latter move and help their erstwhile ally build more resilient institutions. Finally, we turned to the Islamic Republic of Iran. In a roundtable meeting, Hoover fellow Abbas Milani and Stanford University's Roya Pakzad described the Ali Khamenei regime as "tactically nimble" and good at mitigating short-term issues as they arise but "strategically foolish," unable or unwilling to mobilize the vast potential of the Iranian people. Iran currently faces a demographic dividend—a one-time opportunity for an economic boost based on its increased working-age population relative to the number of very old or young dependents. As this dividend passes unrealized and its women's liberation movement expands, the Islamic Republic faces an unstable future.

What about **Europe**? After two destructive wars, Europe flourished in the latter half of the twentieth century and, following the collapse of communism in the East, grew into a powerful economic zone by the early twenty-first century. But the combined forces of demography, advancing technologies, and the information and communications revolution are rapidly changing the continent. Its society is aging while mass migration compounds the challenge of governing over its steadily diversifying citizenry; its institutions face internal and external pressure; and its economic dynamism and primacy appear at risk. Will new means of production catalyze Europe's core manufacturing capacity? Will European leaders master external challenges from the United States, China, Russia, the Middle East, and Africa? And what of the European Union and other institutions, which must contend with new political and populist pressures? It's a mixed bag: the continent has a strong foundation but faces major new domestic struggles that are likely to occupy its attention.

Journalist and former global head of policy at Google Caroline Atkinson reviewed the regulatory landscape and economic potential of digital technologies in Europe, showing how thus far it has been hampered by its fragmented market, but how the continent—with its extensive broadband infrastructure, engineering prowess, and high-quality services—ultimately appears to have a bright future in the digital world. While Europe may have positive structural indicators in the realm of technologies, demographics suggest a more ominous picture. Christopher Caldwell, a contributing editor at the *Claremont Review of Books*, depicted the two overlapping Europes of today: one globalized, wealthy, and generally urban, the other rural, depopulating, and struggling. With its graying societies, European states must address this internal divide and the pressures of migration, largely from the Middle East and Africa.

Europe also faces economic and political pressures from Russia, China, and even the United States prompted by the information revolution and advancing technologies—particularly artificial intelligence. The Brookings Institution's William Drozdiak argued that European states have fallen behind in these areas, but that the US-European relationship can be key to the continent's resurgence.

Continuing that optimistic spirit, Jens Suedekum, professor at the Düsseldorf Institute for Competition Economics, observed that many in Europe appear to have conceded defeat in the race for technological prowess, but that their pessimism is misplaced. He explained that Europe fared better in the first round of the age of digitalization than popular opinion would suggest. Its engineering and manufacturing expertise, supported by its education and apprenticeship systems, which promote worker flexibility, may help it flourish in the next.

As societies across the developed world grow older and some even shrink, **Africa** stands out as the overwhelming engine of global population and labor-force growth, and the source of much of the world's youth in coming decades. Around the time of postwar African independence, there were more than two Europeans for every African on the globe; today, it is Africans that outnumber Europeans two to one, a ratio that will grow to five to one by midcentury, and ten to one by the turn of the next. By the year 2050, Nigeria alone is projected to have a population larger than that of the United States.

Growing workforces are an opportunity for Africa at a time when few other places in the world enjoy similar expansion; by midcentury sub-Saharan workers alone will outnumber Chinese workers, and African workers will surpass the entire Asian labor force before the end of this century. But across the continent African states also face economic challenges, military conflict, and political instability. Moreover, they will feel the effects of a changing climate and environment. Against this backdrop—and keeping in mind the great diversity of the continent—how will African states and people alike realize the promise of advancing technologies and new means of production to advance their standard of living? And what does the spread of communications via social media and mobile technology mean for the future of governance in each region of the continent? Can Africa transform itself—

with its own innovators, influxes of income via remittances from its diaspora, and foreign assistance—from the global engine of population growth to an engine of economic growth? Or will a failed governance response to these challenges make the world's fastest-growing region into an even-faster-growing source of economic migrants and political refugees? In fact, it's more accurate to acknowledge the substantial diversity in countries across the continent. Some are among the fastest growing in the world today and are taking promising early steps toward navigating through the emerging new world; they will need to guide the many others that are otherwise at risk to founder.

As ably directed by our moderator, former assistant secretary of state for African affairs George Moose, contributors assessed Africa's place in the changing world of international trade and the prospects of technological development on the continent. In surveying Africa, George Mason University's Jack Goldstone described its demographic outlook as the most dynamic in the world. But to realize the full potential of that dynamism without losing its demographic dividend to global out-migration, African and foreign states alike should work together to bring down fertility rates and invest in the continent's youth. Complicating those and other development efforts will be climate change, which will affect Africa acutely. Georgetown University's Mark Giordano and Elisabeth Bassini reviewed what we know so far about the future of climate change in Africa—in short, many of the same environmental problems the region faces today will be magnified. Then they called for more research to understand the expected local impacts across the diverse ecosystems of the continent, and recommended ways to prepare for these looming changes.

Meanwhile, technology offers bright spots across Africa. Los Angeles–based entrepreneur Shivani Siroya outlined how mobile internet is opening doors in Africa, as in other poor parts of the developing world, to the direct provision of low-cost banking and financial products needed to launch and scale small businesses. Technologist and investor André Pienaar sees in Africa's youth the potential for significant economic progress through greater access to mobile internet technology. Recognizing how authoritarian states and other actors have weaponized social media and other communications platforms around the world, both participants argued that good governance and a commitment to transparent institutions will be vital to realizing that potential. And while both inter- and intracontinental trade is weak, as described by longtime African businessman Anthony Carroll, educational and infrastructure improvements, coupled with economic reforms, could open the continent further and lay the foundation for widespread adoption of new technological platforms and tools. Overall, Ambassador Chet Crocker (he, too, a former assistant secretary of state) highlighted the future of governance in Africa. He argued that developing economically and politically inclusive governance institutions, within such a complex national and regional political web, will be crucial to the long-term peace and prosperity of African states.

The chapters that follow take on each of these regions in turn, blending that expert commentary with our own reactions. The results, we think, show very different parts of the world, all waking up to very similar sorts of changes, each of them grasping their way forward. Some already have ideas for how to take advantage of the opportunities such changes present; others are looking for help.

So let's set out.

# CHAPTER 1

# Russia in an Emerging World

Advancing technologies and demographics portend disruption in Russia, as in many other parts of the world, but volatility has been the rule rather than the exception for this historic power. The fall of the Soviet Union left modern Russia in a state of disrepair. Its economy collapsed alongside its government. Its population decreased, and its fertility rate plummeted. To those living in Russia at the time, it was deeply destabilizing.

A decade later, Russia's future looked brighter. Its economy rebounded in the 2000s, driven by the country's oil and gas industry. President Vladimir Putin brought stability back to Moscow. But President Putin began instituting regressive policies, and the country suffered from the financial crisis of 2008 and the collapse of the ruble and of oil prices. The imposition of sanctions in 2014, in response to Russia's aggressive actions in Ukraine, further darkened the country's outlook.

Now Russia faces a bleak demographic future: a shrinking working-age population and an aging society. It has a weak, low-tech, slowly growing economy and is ruled by an autocratic regime. Meanwhile it faces the uncertain effects of a changing climate. But it could also be said that Russians today live better than they have for much of their history and that, compared to the 1990s, Russia is stable.

Former Russian foreign minister Igor Ivanov in his essay prepared for this project wrote that the world is at a "bifurcation point." He argued the existing international system cannot last, and the transition to a new one will be either evolutionary (slow and steady) or revolutionary (swift and painful). How will Russia adapt to the rapidly emerging world, with its changing demographics, its spread of information and communications, emerging technologies such as AI, and new means of producing goods near where they are needed? And what are the implications for US strategy in dealing with Russia?

**FIGURE 4 Russia's life expectancy stagnates.** Despite global improvements, Russia's life expectancy stagnated after 1960 and fell to a low of sixty-five years in 2005. It only recently recovered to the level achieved before the 1991 fall of the Soviet Union.

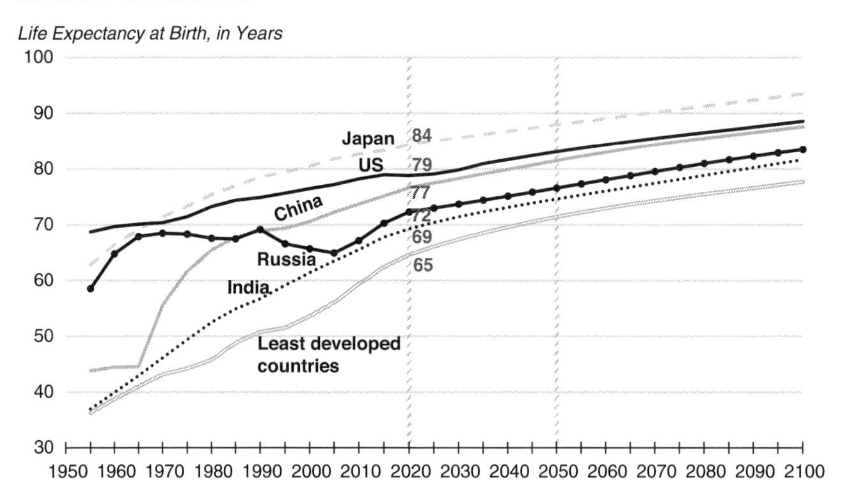


**FIGURE 5 Russian women live eleven years longer.** Life expectancy for Russian men recently improved to sixty-seven years from a low of fifty-nine; women are still expected to live eleven years longer.

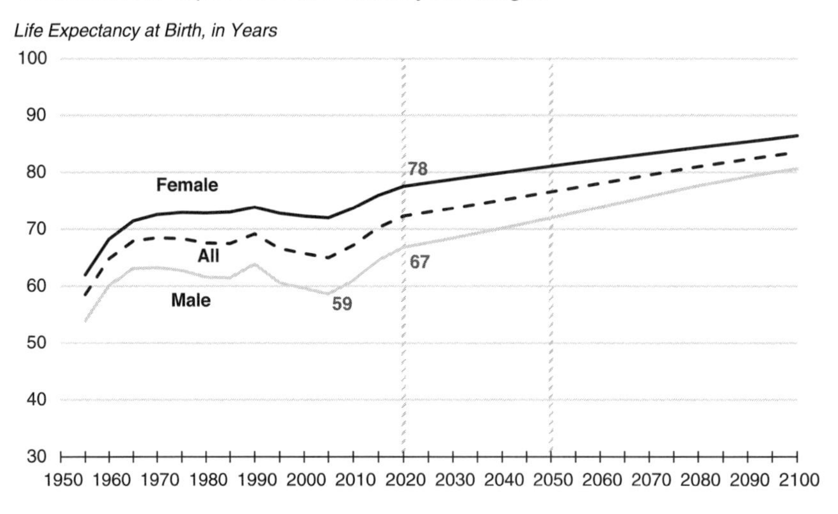


Data Source: United Nations, *World Population Prospects: The 2019 Revision,* medium variant. Demographic analysis by Adele Hayutin, 2020.

The papers prepared for our project roundtable described two competing impulses within Russia's government: the desire to ensure stability, and the urge to modernize and grow. Historian Stephen Kotkin explained, "Russian politics is largely a struggle between those who emphasize the need to gird against perceived threats and prioritize 'stability,' versus those who prioritize development, between those who see the outside world as almost exclusively menace and those who see it as primarily opportunity." In other words, the security services desire security while the finance and economic ministries advocate for development. The security services usually win.

It is important to recall that President Putin sits atop a fractious and volatile government, especially at the elite level. He does not have unitary authority over the state—a large portion of his edicts and policies are not implemented. Some high-level figures, such as former finance minister Alexei Kudrin, urge reforms to address the country's demographic and technological outlook. Others may support the basic structure of the existing Russian regime and its policies but object to Putin's personal leadership of it.

However, President Putin tends to back his security services. Former US ambassador to Russia Michael McFaul wrote that "Putin values control over innovation; vertical instead of horizontal arrangements." While Putin speaks of the need to innovate, Russian policy does not back his words, as David Holloway explained.

If Russia will trend toward stability, what role can it play in the emerging world? What is the Putin administration doing to mitigate the challenges of demographic changes and new technologies? Can Russia have both stability and growth?

## Demographics

Russia's working-age population is steadily declining and will continue to fall for decades. As a result of high mortality rates and an echo of the steep post-USSR fertility drop, Russia is losing a million workers a year. It is also experiencing "brain drain": many of those leaving are young and well educated, while many of the immigrants into Russia are low skilled. As Russia loses the young and educated, it loses scientific expertise. In their absence, it moves in the direction of a petrostate. It is notable, for example, how many Russians work in Silicon Valley or in the flourishing high-tech sector in Israel. Other countries are reaping the benefits of the historic Russian talent for science and mathematics, suggesting that Russia's problems are rooted in its policies and institutions, not its human capital.

While Russia's population ages, much of the developing world is growing more populous. Much of Russia looks warily upon this "youth bulge"—the glut of young people emerging primarily in the southern hemisphere—seeing a mass of potential extremists or a potential dilution of Russia's cultural homogeneity. There appears to be little prospect for immigration to offset Russia's demographic dynamics.

The government has taken steps to address its high mortality rates, low fertility, low life expectancy (see Figure 4), and aging population. Life expectancy at birth for men has climbed to over sixty-six years, but there remains a significant gender disparity; female life expectancy is over ten years higher (Figure 5). An anti-smoking campaign increased public-health awareness and banned smoking in public places, but the next phase—increasing taxes on tobacco products—has stalled. Pro-natal policies implemented in the late 2000s may have helped raise fertility rates. But notably for the prospect of staving off the decline of available workers, the government's proposal to raise the pension age was met with protests across the country. The final decision on raising the retirement age for women represents a compromise.

## Advanced Technology

In principle, advanced technology could help offset Russia's loss of workers by boosting productivity. Moreover, if Russia were to incorporate more advanced manufacturing technologies and robotics, its shrinking workforce might reveal itself as a blessing rather than a curse. The country might sidestep the potential employment disruption such technologies may cause. But the Putin administration chooses to focus on stability and security rather than on adopting the policies and making the investments that would foster development and deployment of advanced technologies on a large scale into the economy.

Recent efforts to build a more dynamic science and technology industry have had mixed results, as seen in its Skolkovo Innovation Center. Although the effort yielded new educational institutions, Skolkovo has failed to blossom into a sort of Silicon Valley or MIT of Russia, despite major investment. However, Russia can count some successes; Yandex, for example, is the lone search engine in the world to compete with Google without substantial government support.

Russian policy choices have hindered the development of the high-tech sector. The government has followed the traditional Russian top-down approach—as opposed to the more organic, bottom-up approach seen in Silicon Valley. The industry suffers from overregulation and pressure from security agencies. The Putin regime has undermined property rights, failed to improve the rule of law (which might have included, for example, a stronger court system or the development of independent government agencies), moved away from democratic institutions, and incurred the wrath of the international system through its foreign policy. Foreign investors do not want to invest in Russian technology—unlike in the oil and gas industry, American tech investors and companies feel no obligation to have a presence in Russia—and even Russians themselves do not want to invest there. All told, Russia would seem to lack the institutions, investment climate, culture of entrepreneurship, and rule of law conducive to a vibrant commercial technology sector.

Instead, high-tech investment centers on military technology. In the United States and China, commercial industries and the military work together to strengthen each country's technological base. Not so in Russia. It is not clear that investing in military technology, absent a parallel supporting and mutually reinforcing commercial industry, will allow Russia to compete with the United States or China. Although Russia and China have explored military technology cooperation, Russia fears becoming a junior partner—or worse yet, a client state—to China. China arguably represents the most serious long-term security threat to the Russian state.

But Russia's military investments have given it effective, asymmetric capabilities, including high-end air and missile defense, cyber capabilities, long-range artillery, and autonomous weapons—to say nothing of its nuclear arsenal. As during the Cold War, Russia will continue to compete with the United States in priority areas, stealing, purchasing, or developing those technologies it views as necessary. And those technologies fostered by the military may yet yield civilian benefits; after the fall of the Soviet Union, for example, the majority of high-quality civilian goods came from the military-industrial complex. Another problem is that these new technologies, armed drones and cybertech in particular, lower the threshold for conflict. At the extreme end of the spectrum, cyber interference in nuclear command and control and early-warning systems could potentially lead to the use of nuclear weapons. All of our project contributors spoke of the importance of renewed communications between the United States and Russia, particularly between the two militaries and between technical experts, to reduce the risk of conflict resulting from misperception and miscommunication. The United States and Russia should reconsider cooperative measures to address common security threats, beginning with discrete, accessible steps.

In sum, Vladimir Putin's rhetoric does not match his actions. He speaks of the need to innovate, saying that a country that rules in AI will rule the future, but has not implemented policies to become such a country. To the contrary, his faith in central planning and distrust of private initiative—manifest in Russia's top-down approach—will likely prevent a commercial high-tech sector from flourishing in Russia, and as a consequence Russia will have difficulty competing with the United States and China in military technology.

## Information and Communications Revolution

Technological developments that might undermine nondemocratic regimes can empower those regimes if they master the technologies, and Russia has done so. It has harnessed the deep digital connectedness of our era to promote nativism and the Putin regime's goals. Traditional media are largely under government control in Russia, and there are growing restrictions on the internet. Three-quarters of all Russians have internet access. They are linked to each other and to people outside Russia. That connection informs and empowers individuals, and to some extent allows them to organize, but on balance Russian authorities and security services use these tools to greater effect for surveillance and repression.

Russia has conducted cyberattacks of various kinds—including interfering in elections—against the United States, European countries (especially the Baltics and Ukraine), and Georgia. As discussed above regarding military technology, Russian investment in cyber capabilities has yielded a significant offset capability—a step-change in skills to which US business and government institutions should respond first with a stronger defense—against those threats that can be defended against. For those threats that cannot be satisfactorily defended against, the US government should build deterrence by establishing systems to publicly report presumptive incoming cyberattacks in real time, and by defining thresholds for acceptable cyber behaviors—and intended responses for when those are breached.

## The Economy

The Putin regime's commitment to stability at the cost of economic and technological development inhibits growth, as discussed previously. Substantial reforms to the rule of law, property rights, and the judiciary are necessary to promote investment.

Russia's GDP is lower than that of Italy, and it has a per-capita GDP on par with Portugal's. State enterprises represent some 60 percent of the economy, and the country relies heavily on hydrocarbon exports, which account for more than half of all government revenue. The demographic outlook outlined here also constrains Russia's economy. As it loses workers and highly educated young people, Russia appears poised for no more than 1 percent annual GDP growth for years to come, of which technological change might be expected to contribute half.

But as is often the case with Russia, it is important to ask: compared to what? Russia's economy may appear weak, but it seems sturdy compared to what it was in the recent past. International sanctions have hurt its economy, but President Putin enjoys popular support.

Authoritarian regimes, such as Russia's, depend on cash flow—revenue to the government—for their legitimacy and survival. They tend to thrive when given resources to guarantee steady revenue. If the global energy market moves to a low-carbon future, Russia could seek to leverage its natural gas reserves as a bridge fuel and lean on its zero-carbon civilian nuclear power technologies—the two primary areas of strength in Russia's energy sector. If hydrocarbon demand starts to wane earlier than expected, President Putin may find himself in need of new cash sources. He may need to depend more on taxation for sustained revenue, with potential consequences for authoritarian rule.

## Climate Change

Traditionally, Russian leaders have put a positive spin on a warming climate. Its effects could, for example, open access to the resource-rich Arctic and allow for agricultural productivity growth. More recent analyses have cast doubt on this rosy assessment, flagging the potential for new natural disasters, alterations in disease vectors affecting public health, and the degradation of existing infrastructure. A changing climate brings widespread costs and demands adaptation. Will the Russian state have the resources and institutions to respond? Strategies for common climate challenges—for example, genetic engineering of new drought- and heat-tolerant plants, better public-health information sharing, or dealing with an opening Arctic—are promising substantive areas for US-Russian interaction during a period of sensitive relations.

Our discussions set out to understand Russia's participation in the emerging new world. In some respects, Russia appears to lack the basis for a large role: it has institutional obstacles to commercial technological development; it has an aging society with low fertility; it is losing many of its best and brightest; and it continually antagonizes foreign powers through cyber malfeasance. It could wind up increasing its dependence on China in its decline. However, Russia has always been a paradoxical country, a nation seemingly perpetually on the wane and yet a permanent fixture in geopolitics.

The volatility of the post–Soviet Union era still impacts domestic politics today. The stability President Putin has brought is highly valued; although public opinion polls in Russia can be suspect, his popularity seemed to soar after the 2014 intervention in Ukraine and has remained high. Yet the regime appears unstable in the longer term. It struggles to implement domestic policies and faced opposition in its major effort to address the problems brought on by its adverse demographics—that is, raising the retirement age. Beyond its borders, Russia tends to emphasize the use of broad multinational institutions, in particular the UN, to address matters of international affairs, preferring institutions that operate only by consensus, where it can control the outcome, and opposing "coalitions of the willing."

Russia generates the economic outlook of a middle power but acts like a great one, and aspires to be greater still. And behind all these issues sits Russia's nuclear stockpile, the largest in the world alongside the United States'.

The United States should work to reopen lines of communication and cooperation with Russia. The way forward is not a grand bargain but concrete steps to build the trust that will be necessary for more consequential steps. Nongovernmental relationships should play a role. Scientific and researcher exchanges and student programs could help the two countries navigate the future of new, developing technologies, such as artificial intelligence. Track II diplomacy—or back-channel diplomacy, which brings together prominent nongovernmental discussants from across countries with the expectation that they may nonetheless be able to influence or otherwise represent domestic thinking—and similar programs, which flourished just a decade ago, could establish open communication, while military-to-military exchanges would lessen the risk of catastrophe.

The Russian system is not conducive to sustainable technological development. Its population is getting older, and it is hemorrhaging talent. Its hydrocarbon-driven economy faces an uncertain future. Russia appears headed toward a significant decline, but it has a long history of mastering its circumstances. Despite its underdeveloped high-tech industry, Russia finds a way—as it did during the Soviet era—to compete with the United States in areas that its government considers priorities. It tends to think not in terms of costs and efficiency, as we so often do, but in terms of objectives. That approach has allowed it to achieve its highest-priority objectives.

When the West looks at Russia, it sees a nation in decline. But when Russia looks back, it sees a West in decline, which the Putin administration strives to outlast. Here again the question of "compared to what" arises: Russia compares itself to the United States, to the West, and to China and seeks to survive as a great power.

Russia is a major power, armed with the most dangerous weapons on earth. It will always be important, so the United States must figure out how to work with Russia constructively. It has been done before, and it can be done today even in a new and changing world.

For more:

- "Emerging Technologies and Their Impact on International Relations and Global Security" by Ivan V. Danilin
- "Russia and the Solecism of Power" by David Holloway
- "New Challenges in Global Politics: A Russian Perspective" by Igor Ivanov
- "Technology and Governance in Russia: Possibilities" by Stephen Kotkin
- "The Influence of Current Demographic Processes on International Relations and International Security: The Russian Take" by Anatoly Vishnevsky
- "The Missed Opportunity of Technological Breakthrough in Putin's Russia" by Michael McFaul

*https://www.hoover.org/publications/governance-emerging-new-world/issue-118*

**From "Russia and the Solecism of Power"**
*by David Holloway*

Nearly every discussion about Russia raises three questions: Who is to blame? What is to be done? And where is Russia heading? This paper focuses on the third question, though the other two cannot be ignored entirely. . . .

The number of multifunction robots per 10,000 workers in the economy has been used as an index of the degree to which manufacturing industry in a country is automated. The world average in 2016 was 74. The highest number was 631 for South Korea. The United States had 189 multifunction robots per 10,000 workers, while Russia had three. . . .

[Alexei] Kudrin pointed to other indicators of technological backwardness. In a long-term strategy drawn up by the government in 2007, it was planned that the number of enterprises engaged in innovation should rise to 40–50 percent of the total by 2020; by 2014 the percentage was 9.9, compared with 8.5 in 2007. Similarly, the share of innovative production was to rise to 20–25 percent from 5.5 percent in 2007, but in 2014 the figure was 8.7 percent. In Kudrin's words, "We set goals but we don't advance toward them." Another important indicator is expenditure on science (state and private). In 2007 it was 1.12 percent of GDP and supposed to rise to 3 percent by 2020, but in 2014 it was still only 1.13 percent. There has been, in other words, a significant and consistent gap between purposeful rhetoric and practical results in this area.

CHAPTER 2

# China in an Emerging World

Aspirational goal setting has been a motivating form of governance for the Chinese Communist Party (CCP). In a 1957 international meeting of communist leaders in Moscow, Nikita Khrushchev proposed that the Soviet Union catch up with US industrial output within fifteen years; Mao Zedong countered, in turn, that within the same time frame China would not just catch up with but surpass the United Kingdom. The result was the disastrous Great Leap Forward and its resulting famine.

But the CCP has also adopted pragmatic frameworks. For example, the founder of modern China's reform movement, Deng Xiaoping, colorfully described the country's process of reform and opening as an incremental "crossing the river by feeling the stones." And Deng's legacy has been extremely fruitful. China's GDP has grown from just 11 percent of that of the United States in 1997 to 63 percent just two decades later. Since the 1980s, eight hundred million people have lifted themselves out of poverty in China, within a population that has gone from overwhelmingly rural and poor to mostly urban and medium income.

Today, a combination of changes that are larger than any single leader or development plan have set up new challenges that lay ahead for China—some of which are unprecedented both in China's own modern history and in all of human history.

One known crosscurrent is demographic, already reported on at length in this volume. The size of China's working-age population has for decades been growing at an average annual rate of 1.8 percent. But China's workforce has peaked and begun to decline in number, and by 2040 it will be falling by 1 percent annually.

Governance is another challenge. The party has taken control of new information and communications technologies both to monitor the views of citizens

**FIGURE 6 Chinese family size dropped sharply.** Births per woman declined sharply even before the one-child policy was implemented. Over the same period, China's life expectancy rose steeply. These exceptionally fast changes transformed the age mix.

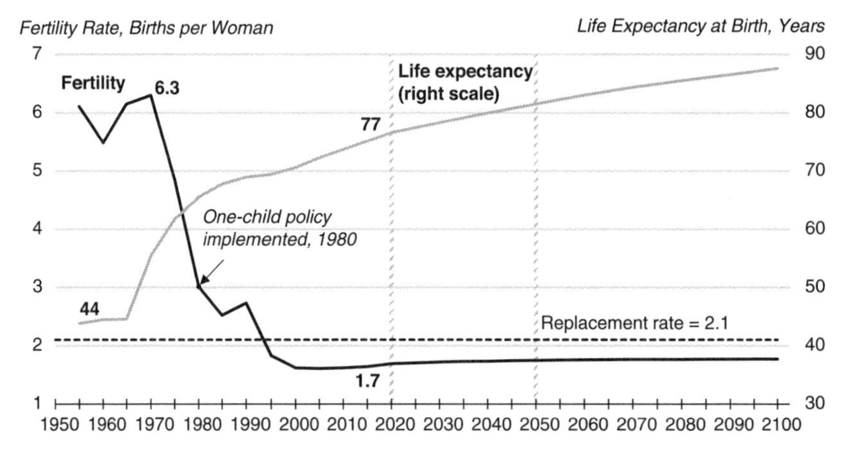


**FIGURE 7 China's workforce shrinks.** After benefiting from a steep increase in its workforce, China now faces a continued decline in working-age population and a doubling of people past working age to nearly 400 million, or 26 percent of the total population, by midcentury.

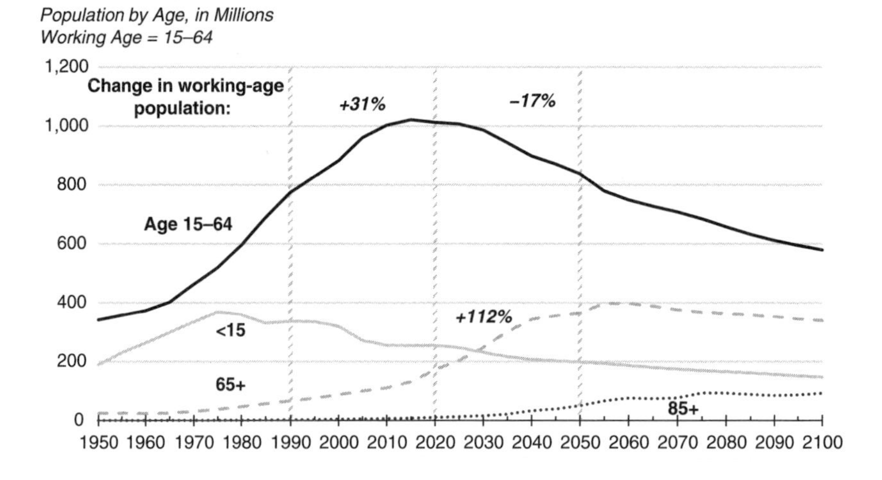


Data Source: United Nations, *World Population Prospects: The 2019 Revision,* medium variant. Demographic analysis by Adele Hayutin, 2020.

and, to some extent, to respond to their concerns. Moreover, it uses these technologies as a means of censorship and surveillance to enforce authoritarian rule. In some ways the party's responsiveness has enhanced the social contract; at the same time it has enabled more complete repression of the citizenry's ability to mobilize and take action. The contradiction between economic flexibility and rigid single-party political control may loom large in China's future.

China's growing economic weight and military capability are also changing the nature of how it balances with other global powers. As China increasingly "goes out" into the world, to borrow a Chinese phrasing, how will those strengths develop?

Meanwhile, environmental degradation of its air and water continues to vex social-satisfaction levels and act as a shadow tax on health and economic well-being. Even in a modern, urbanized China, this gap in environmental performance remains one of the starkest differences as compared with advanced countries such as the United States.

Trying to grasp what these challenges mean for China into the future—and therefore how the United States should position itself—raises the following question: does China's prominence in twenty-first-century technologies such as artificial intelligence, advanced manufacturing, and information technologies now provide an auspicious chance for it to overcome the demographic, environmental, and other headwinds it faces and join the ranks alongside other global advanced economies? Similar challenges have historically hobbled the development of middle-income countries—Mexico, for example, as described later in this book, ran out of its own demographic workforce dividend without breaking out from the ranks of the upper-middle-income countries—but they lacked access to these promising new tool sets.

Our roundtable discussions on China for this project illuminated this question. Experts argued that the challenges facing China are profound, but the regime's adaptation of recent technological innovations has created a stronger and more responsive authoritarian state that will likely pose challenges for the United States in many dimensions.

## Demographics

China's period of reform and opening coincided with a demographic dividend in which its working-age cohort increased its share in the overall population relative to dependents such as children and the elderly. Together with a large, productivity-boosting rural-to-urban migration and commensurate education improvements, these forces helped underpin four decades of 10 percent annual economic growth. While we cannot predict the future of technological progress or political reforms, absent a catastrophic event, China's demographic future through 2040 is essentially set. Decades of low fertility and resulting small family sizes—following the sharp decline of the late 1970s and 1980s (see Figure 6)—means that China's working-age population (ages fifteen to sixty-four), which peaked just before 2015, is now shrinking, and it will continue to shrink with growing speed (Figure 7). Specifically, the pool of potential young workers (ages fifteen to twenty-nine), who are the best educated, most tech savvy, and most flexible in terms of working arrangements, is dwindling sharply—it is also skewed male due to a bias against female children during the (now abandoned) one-child policy era. And the ages-thirty-to-forty-nine cohort—which has been observed historically as a peak age for innovation and high intellectual achievement—is also on the wane.

Meanwhile, the fifty-to-sixty-four-year-old cohort—the least educated and healthy of China's workforce—is growing, but it will start to shrink in number by 2040. And with rising longevity, the elderly population over sixty-five is exploding. The pool of dependent seniors is growing by 3.7 percent per year, from 135 million in 2015 to 340 million in 2040, at which point it will constitute 22 percent of the population, making China a "super-aged society." These headwinds imply that the coming generation will not see the phenomenal economic rise of the last without radical productivity-improving interventions: as demographer Nicholas Eberstadt observed, "The demographic dividend has already been cashed." These new pressures also could lead to social and political instability.

For example, how will China provide for the immense population of future seniors? Other countries, including the United States, face similar challenges, but with a considerably smaller scale and intensity. Today's rudimentary and uneven pension and health system does not cover many Chinese seniors, yet the current unfunded pension and health liabilities are already comparable in scale to the country's annual GDP. Traditionally, the elderly would rely on their children and families to support them. However, the small size of most Chinese families—a product of low fertility, whether forced or otherwise—limits their ability to perform that traditional caretaker role. The elderly will furthermore be concentrated in rural areas, where social services and other resources are weakest. Rural China, in particular, is set to become one of the grayest populations ever observed in history. Caring for the elderly could lead to substantially increased welfare spending going forward, reducing resources available for other productive investments or geopolitical priorities such as foreign lending or military modernization.

Meanwhile, China continues to rely on domestic migration from rural to urban areas to maintain the workforce, both from villages to midsize cities within provinces and from the inner hinterlands to coastal megacities across provinces. Migrants—more than 40 percent of the urban population—are engines of economic growth. But they remain largely second-class citizens without urban residency status. Granting entitlements such as education rights for children of migrants could also redirect city budgets that have in recent years otherwise been available for large spending projects on infrastructure or security regimes.

Finally, there is the impact of extremely low fertility rates on social structures. As early as 1990, for example, four-fifths of children born in Shanghai had no siblings, and those children themselves have now produced their own "only children." This has led to the creation of an inverted "new family type" in China without any brothers or sisters or extended relatives—only ancestors. It has also led to a gender imbalance: by 2030, 20 percent of Chinese men in their early forties are expected to never have been married, up from just 4 percent in 2000. This new dynamic causes parents to devote immense resources to their child's upbringing but also could generate a growing sense of social risk aversion. Parents, for example, may be less willing to support their only children in taking personal or employment risks—or even more so military engagements, where the potential loss of sole lineage-bearers becomes intolerable. The Chinese government is aware of these demographic challenges. It continues to emphasize urbanization, along with education, as an engine for productivity growth and has taken more radical steps to try to affect culture: the wave of thousands of state-sponsored tech incubators in recent years has shifted modern Chinese attitudes toward embracing entrepreneurial risk taking.

But the scale of these overall changes suggests that demographics will continue, in Eberstadt's words, "to bound the realm of the possible." Growing acceptance of immigration in a historically closed society may help China attract workforce participants of the highest quality, but even large flows from China's smaller neighbors cannot appreciably impact the question of quantity. Meanwhile, any future pro-natal efforts by the state are likely to be as ineffective as they have been in other East Asian countries (Japan has struggled for years to encourage a higher birthrate through generous social support policies like preschool subsidies)—unless China discovers new ways to use twenty-first-century social-monitoring technologies to incentivize or coerce its citizens' behavior to these ends.

## National-Security Implications of Advancing Technology

In the military arena, China sees an opportunity to use twenty-first-century technologies as a route toward overcoming the United States' traditional dominance—one which they see as based on legacy systems potentially vulnerable to warfare of the future, including cyber weapons, artificial intelligence–enhanced information systems, and autonomous platforms.

Military technology analyst Elsa Kania pointed out in her writings for this project that artificial intelligence in particular has emerged as a new focus of competition between the United States and China: prowess in machine learning and big data has strategic significance for both economic development and military modernization. And this competition feeds on itself: current US superiority, demonstrated by computer program AlphaGo's surprise victory over the best human Go players, stimulated China—the overtones of dominance in this traditional game of military strategy did not go unmissed by Chinese observers. The United States' own pursuit of AI for national-security purposes is in turn stimulated in part by China's massive investments in this technology. Both parties believe AI could be revolutionary, and that it could disrupt the military balance. China seeks the "intelligentization" of its military—wherein it uses a variety of large data sources and machine learning to inform operational decision making, enable new capabilities, and even change its force-generation models—in order to augment its forces and become capable of fighting, and winning, on the modern battlefield. At present this is largely an aspirational goal supported by research, development, and verification that will play out for years to come. To do this, China both closely studies US defense innovations and is mobilizing its own domestic firms and universities to support military use of advanced technologies. For example, the People's Liberation Army (PLA) is now supporting a range of projects involving applications of artificial intelligence and related technologies for target recognition, electronic warfare, resilient communications, cybersecurity, and defensive and offensive systems. The PLA is pursuing autonomous vehicles—for air, sea, and land operations—enabled by machine learning and additive manufacturing, including swarming capabilities as an asymmetric counter to US legacy platforms. It is also exploring use of artificial intelligence to support rapid decision making. China already contends with the United States for superiority in the global military unmanned aerial vehicle (UAV) market and is home to the largest manufacturer of commercial drones.

More broadly, it is important to appreciate that the combination of artificial intelligence and cyber warfare could potentially escalate future conflicts. Complex AI systems involved in decision making and operations could produce mistakes, triggering unintended and perhaps escalatory consequences as they do. And going forward, both China and the United States may in fact share similar vulnerabilities as advancing technologies make asymmetric weapons systems across various domains more accessible to less sophisticated players. The United States and China would be wise to continue track II diplomacy and eventually military-to-military dialogues to reduce risk in this area.

## Advancing Technology and the Chinese Economy

In artificial intelligence, China sees an opening to move beyond its fast-follower economy. As technology investor—and renowned inventor of natural language processing technologies himself—Kai-Fu Lee and Silicon Valley–based China observer Matt Sheehan shared, the Chinese government and firms are joining others around the world in exploring the new "plateau" that has been created by recent advances in machine learning and big data. Chinese productivity in the period of reform and opening and rural-to-urban migration has already increased by a magnitude of twenty. As demographic and other economic challenges now loom, one goal is to apply these new technologies throughout many sectors to help extend such growth.

Enthusiasm in the country is heightened by a sense that China may in fact have comparative advantages in the application of AI throughout the economy. Early in the twenty-first century, following decades of research and building on an explosion of digital data and advances in computing power, machines became able to learn from the data, recognize patterns, and predict the best answers, allowing them to take on many tasks such as driving a car, diagnosing a disease, and making a loan. This advancement marked the culmination of a period of great "discovery," which favored the most elite research institutions at universities and global companies, such as those in the United States.

Now, however, the focus of the field may be shifting toward "application" of this technology, which could play to Chinese strengths. Lee argues that this period of implementation favors the engineer, not the scientist: China produces large numbers of well-trained, competent engineers to fill the ranks of start-ups and large companies. And vast amounts of data, the raw material for applying machine-learning algorithms to real-life problems, are simultaneously being generated as Chinese citizens willingly funnel a growing portion of their daily activities through their smartphones, providing Chinese companies with a deep and multidimensional picture of their lives. Chinese legal and interpersonal trust–based transactional norms remain poorly developed compared to the mature social institutions of the United States, so it's actually unsurprising that Chinese consumers might prefer to actually mediate those same interpersonal experiences through objective digital decision-making platforms such as those offered by artificial intelligence.

In short, while the US technology ecosystem excels at innovation—that is, the creation of an original product or service—China's technology ecosystem instead excels at imitating and improving on a successful business model. When a new approach proves worthy, dozens or even hundreds of Chinese start-ups flood in and compete ferociously, exploring hundreds of variations. Few survive, and the process rewards those best at iteration and execution—a process that stems from but ultimately goes beyond simple copying. Meanwhile, China is now more open to accepting that while market exclusion and intellectual-property theft helped it gain a strong digital-technology position in the first place, it should now protect its own newer domestic innovations built on top of this foundation. Chinese firms, having trained through extreme competition at home, seek to beat US rivals in exporting these products to the developing world, especially Southeast Asia and Africa—and extending Chinese global influence in the process.

To take advantage of this perceived opportunity, the central Chinese government has provided signals to local officials encouraging them to support investments in artificial intelligence technologies, through "guiding funds," public projects, and performance evaluations. While some commentators have argued that the United States federal government should respond in kind with its own AI strategy, it is important to keep in mind the fundamental differences and different tools available in China's "totalitarian market" economy and top-down governance system. Even in China, the most impactful applications of artificial intelligence today have largely come from private technology firms at arm's length from the central government and largely have been pursued for commercial ends. Kai-Fu Lee observed that today the world's most valuable speech-recognition, machine-translation, drone, computer-vision, and facial-recognition companies are all Chinese.

Alongside these advancements, many jobs throughout the Chinese economy are vulnerable to disruption. One category is physical work. Nearly half of Chinese workers are on farms or in factories, often performing repetitive tasks that are subject to automation and robotization. Productivity of today's rural workers is particularly poor in China, especially in regions with small plots, lack of investment, and poor logistical infrastructure. Meanwhile, China is by some measures already the world's most competitive manufacturing power. It is also already the world's largest robotics market. But industrial robot adoption levels per worker throughout the Chinese manufacturing sector remain below the global average, and firms have responded by attempting to acquire established overseas vendors and technologies. The pace of disruption felt in both fields will depend on how quickly capital-intensive automation or advanced manufacturing practices can be successfully applied to these tasks.

Another category is white-collar jobs. Here the impact on employment is likely to be varied. Machine learning performs well under certain conditions—abundant data, narrow goals, and clear outcomes. But it struggles with strategic tasks, unclear goals, and creative or social tasks—all areas where many of today's white-collar employees are likely to have more job security. Many care, service, and creative jobs—at both the high and low ends of the economic spectrum—are likely to be secure. Middle-income college-educated office workers, however, who take data and make predictions face risky prospects. The implications for this uneven impact across economic groups will be a challenge that China shares with other modern economies, including the United States.

## Communication Technologies and the State-Citizen Relationship

The US-based communications professor Maria Repnikova observed in our discussions that "much of Chinese life is increasingly virtual," a transition that has not led to democratization of the country but has created an "unprecedented space for public expression."

WeChat, for example, has more than a billion active users. Individuals employ WeChat to connect with each other, receive news, and make mobile payments for almost everything. Through platforms like WeChat or the Weibo microblogging social network, society can now access prominent journalists, lawyers, professors, and intellectuals. Civil society groups collaborate. Digital peer-to-peer financial technologies have gained rapid adoption in a society with a traditionally underdeveloped personal financial system.

As Chinese citizens moved toward digital communications and social media platforms in the first decade of this century, traditional news outlets struggled in China as elsewhere globally—because, of course, news breaks first on the internet. In China, this demise also meant the weakening of conventional propaganda. Unlike before social media, now authorities at times lose control of discussions that start online but sometimes move into real life as well.

This digital communications revolution created a new vibrancy in China, especially since the first movers to occupy this technological white space have generally been private parties who otherwise faced great barriers to success and innovation in state-controlled media. And private Chinese citizens have been reaping the benefits of that, despite using an internet that has been largely, and is increasingly, cut off from access to foreign content or participation by foreign firms.

Meanwhile, Chinese authorities are responding. Xi Jinping's administration has placed the battle for public opinion as a core party objective, adopting sophisticated digital media PR strategies that seek to leverage these technologies to improve operational governance—"responsive authoritarianism"—while nonetheless attempting to strictly control any content that could lead to mobilization or action. This has led to a complex relationship between authorities and individuals.

On the one hand, censorship and surveillance are used to support authoritarian rule: journalists and editors receive censorship instructions, censors weed out negative elements, and commentators—paid or voluntarily nationalistic—

inject pro-regime views. As an explicit prerequisite for doing business, private Chinese internet companies furthermore funnel real-name user information on all users to authorities, allowing them access to vast collections of citizen data. Firms use this to mine personal information and train algorithms for new products and services. Authorities use it to monitor the activities of individuals and to establish "social credit" systems to reward pro-regime behaviors and punish activities contrary to the regime's wishes. Most darkly, China's advanced digital-surveillance capabilities have been utilized to turn entire provinces into near police states in the name of public security: increasingly, private intrusions have been piloted on the daily lives of the Uighur ethnic minority in Northwest China's Xinjiang Autonomous Region, elements of which are then spread to the broader Chinese populace over time.

But authorities also use the internet to identify citizen concerns and solicit feedback—using these digital tools as a new way to boost legitimacy. The CCP has built an immense infrastructure of social media minders at all levels of government to manage and monitor its network platform–based interactions with the public—that is, internet-based forms of communication and social interaction that spill into the offline world too. Official social media mixes the party line with entertainment and advice. Censorship is intentionally incomplete and sporadic, allowing for a publicly visible but nonetheless controlled venting of opinion. Individual expressions of discomfort on topics including corruption, pollution, and inequality are monitored to gauge public sentiment. This is all up to a point: authorities are most sensitive to discussions that could lead citizens to mobilize and take action. Uncontrolled cyber nationalism, especially among youth, is also regarded as potentially destabilizing when it ends up constraining authorities' decision space. For example, during the 2019 Hong Kong protests, hard-liner mainland Chinese netizens, first inflamed by propaganda outlets, later began to own the narrative—making it difficult for mainland leaders to lower the temperature of the Hong Kong conflict through public compromise. Chinese citizens and officials both are adaptive and innovative, resulting in a more politically active society and, to adopt Repnikova's phrase, a more responsive authoritarian state.

Overall, authorities regard China's internet-management strategy as a success, given the threats these technologies looked to pose to the regime in the early 2000s. As with commercial technology firms "going out" to the wider world, authorities also aim to export scaled-down models of Chinese-type "cybersovereignty" to other interested developing nations. This suggests increasingly divergent US and Chinese global models of internet governance, digital technologies, and society-state norms. Against this backdrop, some commentators suggested that US tech firms should consider how engaging in the Chinese market may negatively affect their own global images. Others, however, pointed out that US firms should resist shying away from developing country markets for values-based reasons only to cede that market to players without any Western influence—Facebook, for example, may not have a perfect response that would limit the incitement of violence against Myanmar's Rohingya Muslim minority by users of its platform there, but its continued presence in Myanmar is still preferable to the alternative of Chinese social network alternatives that would rise in its place were it to withdraw.

Ambassador Stapleton Roy observed that China is acutely aware of the heavy price it has paid throughout modern history for its technological backwardness. But today, Chinese authorities see an opportunity to level the playing field. In this round of history, both commercial and state actors are mobilizing to become masters of twenty-first-century technologies, not just to neutralize this historical source of weakness but also to use these technologies as a way to leap over the significant challenges the country now faces. These capabilities are seen as a potentially novel path to overcoming the middle-income trap (whereby weak institutions—government, civil, rule of law, investment, and so forth—prevent developing countries from durably transitioning to the ranks of wealthy ones, despite initially rapid expansion) through continued rapid economic growth until midcentury, while sustaining Chinese-style authoritarianism.

One question is: are Chinese-style central-planning institutions capable of handling this complexity and uncertainty? It is important to understand, however, that while Beijing may set goals for technological achievement, the relatively small central government and party apparatus often rely on distributed, bottom-up regional efforts to implement them. Private tech firms in China develop, say, advanced facial-recognition capabilities because they have a willing municipal public-security bureau buyer—and will also be guided along the way by a compulsory board seat given to a party member. A local cadre grants free land to a local agricultural robotics manufacturer to cultivate a new source of tax revenue—and garners positive career promotional consideration in doing so. A county environmental agency receives central-government funding to hire a full-time social media monitor, using public opinion to prioritize enforcement efforts on the worst polluters—while censoring efforts to plan a citizens' protest. Moreover, as our panelists observed, Beijing's pragmatic streak means that midterm corrections can be made, often through quiet reorientation, when reality is not going to plan.

Despite this, some surprising retrenchment is visible today in Xi Jinping's China. For example, Beijing has returned to debt-based infrastructure spending to counteract economic slowdowns. It has fallen back on new rounds of subsidies to struggling state-owned enterprises rather than letting their more nimble domestic private-sector competitors allocate capital. The urban-rural *hukou* identification system, a vestige of Mao Zedong's authoritarian cruelties, persists as too attractive a tether on society to let go, despite the economic and potential humanitarian benefits of reform.

Moreover, China's economy is probably not as strong as it appears. A recent Brookings Institution paper by researchers at the University of Hong Kong and the University of Chicago estimated that China's annual economic growth rates from 2008 to 2016 had been overstated by as much as 1.7 percent each year, and its investment and savings rates by even more. As the authors explain, China's central government calculates the country's economic performance using data from local officials, who are incentivized to exaggerate their own rates. Other analysts have pointed to economic implications of the country's extensive shadow banking and hidden debts. Such miscounting can add up over time, and it has actually led to the rise of cottage industries using new technology such as commercial satellite imagery and other proxies to arrive at more actionable estimates of Chinese economic activity. Whatever the true figures, this points to the importance of looking for Chinese strengths and weaknesses as they wax and wane through stages over time, and not through a narrative of predetermination.

All told, China faces serious countervailing trends: technological dynamism and a novel model of agile governance weighed against demographic decline, authoritarian rule, and environmental challenges. As it seeks to escape the middle-income trap, will artificial intelligence and other advanced technologies spur sufficient productivity growth—and do so quickly enough—to counteract the loss of labor supply? Will they provide a great leap forward and make China a modern "moderately well-off society," despite the commensurate social disruption risks, such as job losses—or will they rather serve as stepping-stones toward more incremental change? Chinese domestic perceptions of progress here are likely to affect geopolitical calculations as well: the timing of and priority placed on China's goal of forcefully "reunifying" a de facto independent Taiwan, through military invasion if necessary; or the handoff between Deng Xiaoping's aphorism of "biding time" by keeping a low foreign profile versus Xi Jinping's consciously outward-projecting "One Belt, One Road" diplomacy initiative of foreign lending, company acquisition, and infrastructure development. Overestimation of technological progress could lead to miscalculations of ambition and capability.

In particular, China's emerging means of governance through its use of these technologies—responsive authoritarianism—brings new wrinkles to the old image of a centralized state. Some roundtable participants speculated that, going forward, the government may lean harder on these tools for tracking citizens' activities and coercing or incentivizing behavior. Might this approach be extended to address China's demographic challenges, perhaps to encourage higher fertility, for example, by crediting those who have more than two children while punishing those with fewer?

Beyond its borders, China may seek to export this governance model (which it views as legitimacy enhancing) to willing developing states as an alternative to the West's own export of liberal democracy. The spread of Chinese-style authoritarianism may therefore pose an ideological challenge for the United States and its traditional values of a free internet, free expression, privacy, and democratic governance.

China's technological prowess and progress challenge US supremacy in this area as well. It may be tempting to try to hinder Chinese technological efforts in a bid to maintain superiority or in justifiable grievance for China's past failures to abide by international rules of trade or statesmanship. In the long term, as we deal with China as both a strategic competitor and a major trading partner, much will depend on how the United States addresses its own headwinds on these same matters of demography, technology, and governance over diversity. There is an interplay.

For the United States, it is important that we do not try to "out-China" China. Rather, we should focus on bolstering our own strengths. First, the US government should continue to make the global case for liberal democratic values, which retain universal appeal if not universal applicability. It should also support the ability of the US private sector and civil society to do the same through cultural and human outreach; this is an area in which the United States is an undisputed leader and China shows few prospects of catching up.

More concretely, US values can be applied toward updating ideal models of internet governance and business too. In the early years of Web 2.0, US digital firms decried then-nascent efforts by China to first censor and later close off its internet market. More recently, the United States focused on European efforts to fine, tax, and closely regulate US internet businesses. As US society now reexamines some of its own attitudes about domestic digital communications, it should do so with an eye toward ensuring both the competitiveness and attractiveness of US internet business and governance models in developing countries, the next wave of global online marketplace growth. Becoming too inward looking at this point in history could have long-term implications for our ability to affect ideals and norms in this increasingly central realm of global influence and culture.

And similarly, as technological advances such as automation change the nature of work in this country, the United States should not forget that, just as after the Second World War, it remains part of the messy world, whether it likes it or not. Lessons learned, positive and negative, from the last three decades of trade liberalization can be applied to technological disruptions—we can do better from experience. The American government need not copy the Chinese state's AI development campaign, for example, but neither should it stand in the way of productive technological development for fear of social impact. Our global competitors with no such compunction may instead end up being the ones doing the disrupting. At minimum, the United States should ensure that it has created the framework to enable private development of twenty-first-century technologies: rule of law, strong infrastructure, minimizing of the costs of doing business, liquid markets, and a predictable regulatory framework.

Technological advances will also color the strategic competition between China and the United States. Chinese military planners see a future of warfare enabled by data and computational capacity, and are pursuing a host of AI-enabled military capabilities—an effort made possible in part by the substantial, if compulsory, civil-military fusion of technology companies. The full US response to China's military expansion is a topic for another time, but, as in the realm of internet governance, the United States cannot emulate the

Chinese approach here, nor should it. Instead, the United States must work to secure and sustain its superior technology, committing the resources and infrastructure necessary to expand and capitalize on research into the military applications of AI. At the same time, it must remain cognizant of the risks to strategic stability posed by AI-enabled early-warning and command-and-control systems.

Finally, there is the environment. The United States is fortunate to possess a variety of excellent natural resources and to have had a century of experience in developing the complex institutions necessary to preserve human health and environmental quality while sustaining a robust industrial base. Chinese (and other) visitors to the United States routinely remark at how enough money can buy most material aspects of American life in modern cosmopolitan cities such as Shanghai or Beijing—but it cannot purchase clean air, water, and food. In China, more than 1.5 million deaths each year have been attributed to air pollution, and drought pressures, already severe, are expected to grow. As the United States justifiably seeks to maintain global economic competitiveness while staring down potentially daunting new environmental and resource challenges such as climate change, it should play to its own ideals and be careful not to accidentally let go of the things that make America great today.

For more:

- "China's Demographic Prospects to 2040: Opportunities, Constraints, Potential Policy Responses" by Nicholas Eberstadt
- "The AI Titans' Security Dilemmas" by Elsa Kania
- "China's Rise in Artificial Intelligence: Ingredients and Economic Implications" by Kai-Fu Lee and Matt Sheehan
- "How Chinese Authorities and Individuals Use the Internet" by Maria Repnikova
- "China's Development Challenge" by J. Stapleton Roy

*https://www.hoover.org/publications/governance-emerging-new-world/fall-series-issue-218*

## From "China's Demographic Prospects to 2040: Opportunities, Constraints, Potential Policy Responses"

*by Nicholas Eberstadt*

Today, "only children" form a majority of urban China's (within the legal *hukou* system) population under thirty-five years of age—and a super-majority of the under-thirty-five population in the country's big cities. This means we are starting to see the rise of a new family type in China: only children begotten by only children—boys and girls with no siblings, cousins, uncles, or aunts, only ancestors and (perhaps eventually) descendants. For this new family type, the traditional extended family has essentially collapsed. This new family type is now beginning to account for a sizable fraction of urban China's children—very possibly, an outright majority in the country's economic and political nerve centers (Shanghai and Beijing) and in other cities of size as well. But even in places where the emerging new family type does not dominate, in the rising generation who will be the parents of 2040, the extended family and its kinship networks are being dramatically compressed by long-term sub-replacement fertility. . . .

What happens though when there is no living son? We are about to find out, and big time: back-of-the-envelope calculations suggest that the proportion of Chinese women sixty years of age with no male child will have risen from 7 percent in the early 1990s to 30 percent or more for post-2025 China. Dutiful daughters may of course step in, but their loyalty, attention, and resources may be all-too-frequently divided, inadequately, between two sets of aging parents.

All of this, however, presupposes that two-and-a-half millennia of Confucian values will inform the behavior of adult children toward their elderly parents in the generation to come. That means taking the near-universal continuation of filial piety for granted. Such devotion might have been easier when the elders were scarce and the children were plentiful; tomorrow those tables will be turned. Beijing has already begun to lay down markers here, criminalizing nonsupport (and even nonvisiting) of parents in 2013. Why, we may wonder, do authorities feel such laws to be necessary?

### From "China's Rise in Artificial Intelligence: Ingredients and Economic Implications"

*by Kai-Fu Lee and Matt Sheehan*

It's useful to introduce an analytical framework popularized by Peter Thiel: the difference between "0 to 1" and "1 to n" innovation. Zero-to-1 innovation describes the process of creating original and radically new products or services. By contrast, 1-to-n innovation involves scaling up and iteratively improving an existing offering. This clean dichotomy is inherently reductionist: no new products truly begin from "0," and moving from "1 to n" is not the linear process implied by the title. But it does offer a conceptual frame that is useful for understanding the unique cultural undercurrents of China and Silicon Valley.

Silicon Valley (and the US tech ecosystem more broadly) both prides itself on and excels at 0-to-1 innovation. In US tech circles, there is great prestige attached to outside-the-box thinking, and significant levels of stigma for those who merely imitate existing models. As a result, the US ecosystem carries a more significant first-mover advantage, allowing the pioneers of a model to patiently harvest the low-hanging fruit born out of their original idea.

China's technology ecosystem, by contrast, tends to excel at the 1-to-n part of the innovation equation. Chinese tech entrepreneurs are far more cautious when it comes to experimenting with radically new ideas, but they have no hesitation when it comes to imitating and improving on a successful business model. The reasons for this are complex, ranging from millennia-old cultural traditions to the breakneck pace of economic development in recent decades. But the results are clear: when a new technology or business model is proven to work, dozens or even hundreds of Chinese start-ups flood into that industry and compete ferociously for dominance.

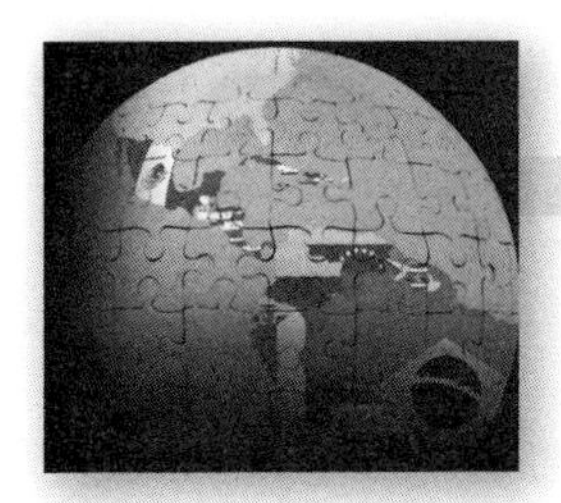

CHAPTER 3

# Latin America in an Emerging World

Our takeaway from a study of Latin America in an emerging new world is an understanding of a region showing gradual—and fragile—economic, social, and governance progress *on average*, but with significant heterogeneity lying beneath, both within and across individual countries. For example, while Mexican manufacturers are by some counts already more robotized—and therefore more ready for future disruptions—than those in the United States, citizens in some rural parts of the country live with few opportunities, in conditions more closely resembling those of sub-Saharan Africa in areas such as health care, energy, and the rule of law. And while Uruguay, Costa Rica, and (until recently) Chile show a consistent trend of stability and growth, with rising quality of life, their next-door neighbor Brazil, once a developing-economy powerhouse, has fallen to below-investment grade and suffers rising drug violence. Colombia, post–peace agreement, has seen a historically remarkable economic and governance turnaround, while the economy in nearby Venezuela contracted by 18 percent last year, casting refugees across the continent.

We define something as fragile if it can break down easily or quickly. Even as we look to long-term forces that may shape the future of this region, then, its persistent fragility appears set to define its day-to-day realities, country by country, and vote by vote.

Meanwhile, though specific demographic pictures vary across the region, one consistent observation is that the future composition of each country's population (size of labor force, age, fertility) and ensuing migration pressures will not look like they did in the past. These longer-term facts have not been a focus of governments in a corner of the world that has seen an average of thirteen new constitutions written per country over the past two centuries.

**FIGURE 8 Northern Triangle out-migration peaked in 2000.** While US crossings have become more visible in recent years, net outflows from the Northern Triangle of Central America peaked in 2000. Future out-migration will depend on economic conditions and border policies.

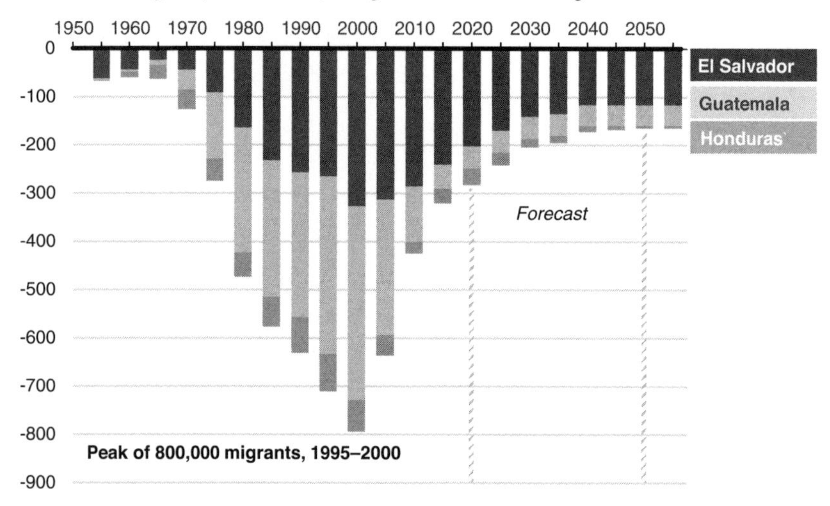


**FIGURE 9 Latin America's fragile democracies face shrinking workforces.** Brazil's workforce is projected to drop sharply after 2040, and Mexico's workforce will peak around 2050. In contrast, Guatemala's working-age population will grow through 2070, nearly doubling from today.

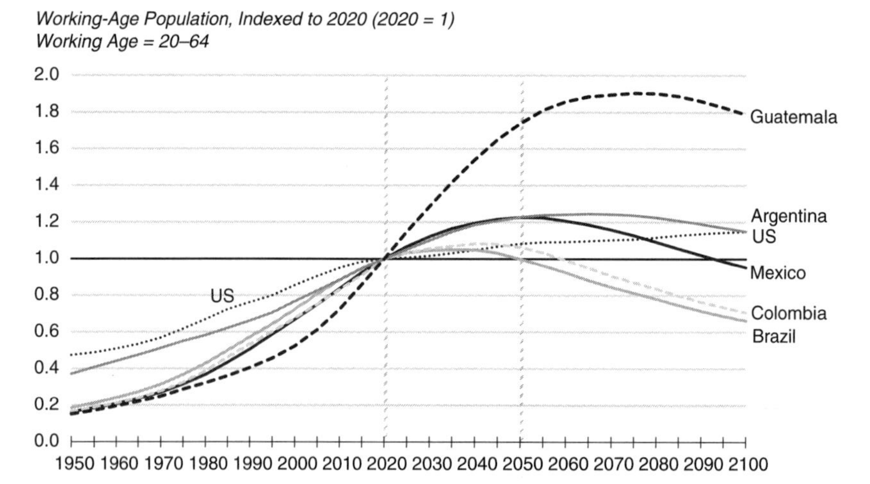


Data Source: United Nations, *World Population Prospects: The 2019 Revision*, medium variant. Demographic analysis by Adele Hayutin, 2020.

Latin America has promise, and clear bright spots, but the status quo appears fraught given the changes that are already arriving across the region. In the near term, Latin America appears set to be a technology taker, adapting externally developed and managed platforms and systems to its local needs. There are risks to existing blue- and white-collar jobs, along with potential for transitioning citizens out of the informal sector. At the same time, the urbanization of Latin America (80 percent of the population lives in cities) would seem to facilitate further adoption of value-adding digital technologies into governance and everyday lives. We think this argues for a nuanced approach to twenty-first-century technologies in the region.

In regions and sectors where existing institutions are generally strong or at least improving, gradual implementations into areas including digital governance, e-health, education, and manufacturing are likely to yield positive marginal dividends. Here, a rising middle class is demanding new attention and services that their governments have not traditionally been focused on offering. This is a fertile space for new technologies to take hold. Most recently, in Mexico and Brazil (together representing half the region's population), campaigning politicians have already seized upon social media and new communications platforms with surprising—even alarming—speed. Next, those same tools may also be applied to governance more broadly.

Meanwhile, where today's conditions and institutions are stubbornly poor, for example in the violence-torn Northern Triangle of Central America (NTCA, encompassing El Salvador, Guatemala, and Honduras), more innovative applications of new technologies may be able to circumvent institutional weaknesses to create "pockets of success" that improve the lives of everyday people. Digital network platforms and the firms behind them could stand in for failing tax collectors, schools, even social trust systems. But even they demand a minimum of traditional social services and public infrastructure to help improve the region's productivity—it is difficult to run a business in a country where only one-third of the young finish high school and with just seven feet of road per capita, versus over seventy feet in the United States.

The United States can help in some aspects of this transformation. And engaging in those areas where the United States can be effective shouldn't be seen as taking responsibility for another country's citizens, but as improving the "neighborhood" in a way that serves our own self-interest too. There are US domestic implications to what happens in a fragile Latin America (see Figure 8). But choices made in Mexico, Central America, and South America against this backdrop of global technological change over the coming years also hold broader "great power" geopolitical implications, including growing Chinese economic and political interests, that have gone unappreciated in traditional US narratives on the region. Things can change fast in this neighborhood. We need to pay attention.

## Technology and Demographics in the "Fragile Democracies"

Former Chilean parliamentarian and editorial columnist Ernesto Silva, in his survey of the region in this volume, points out that Latin American economies were growing on average by only 1.2 percent in 2018 (faster if Venezuela is excluded) versus 3.7 percent worldwide. Despite low economic growth rates, however, social indicators show poverty levels have been reduced to below world average, and there is in many of these countries a growing middle class. Across the region, growth of the labor force is moderate, but generally well down from past growth rates, and will soon diminish further. In the key countries of Brazil, Colombia, and Chile, the labor force will begin to decline by 2035 (Figure 9). With this comes a rapidly increasing population of older adults—by 2050, for example, Brazil's population, in terms of median age, will be older than that of the United States.

These trends point to a demographic dividend window that is still open but rapidly closing. Can the region take advantage of that? Labor productivity remains stubbornly low. Public debt is growing and will likely continue to grow absent reforms to social security systems that will come under stress from an aging populace. Alongside that, credit ratings are being reduced, increasing the cost of new infrastructure investment and doing business in general.

Given this, governments and institutions in the region will need to do more with less, especially in countries that have ridden the now-ending commodities boom—the one-time Chinese demand surge, now moderating, for regional primary exports like oil or iron ore, even soybeans. In countries with general stability and a rising middle class, citizens have new demands, and new frustrations. In Chile, the state, for example, is now expected not just to provide universal access to primary education but to improve the quality of secondary and college education too. Youth street protests there through the fall of 2019, sparked by a marginal subway fare increase, can be seen as a boiling over of middle-class angst. Commenters at our roundtable pointed to the Brazilian middle class's (justified) dissatisfaction with their government's recent performance as the underlying driver of populist Jair Bolsonaro's surprise election, which came on the back of 8.7 million Facebook followers and just 18 seconds of television time. Similar strains of malaise are evident in Mexico, where real incomes for college-educated workers have gone down in recent years despite overall economic growth, and where citizens' dissatisfaction with corruption and violence is escalating.

What does technology hold for these relatively stable, if fragile, parts of Latin America, and how can the United States play a part? Access to the internet is growing, and 56 percent of the region's inhabitants were online in 2016, above the global average. But the pace of modernization of state and local government services and business innovation has been slow. We believe that information technology provides opportunities to increase growth and lower dependency on commodity-price cycles. It can reduce the cost and improve the quality of public services, and governments can be restructured to take advantage of these new technologies.

There are a variety of ways that a more aggressive technology uptake in Latin America could help close the gap with the rest of the world. Consider finance, where 50 percent of the population is currently not part of the banking system, and where new forms of money and digital payment platforms offer a significant opportunity. Uptake of digital banking offers the opportunity for electronic commerce, increasing competition and improving quality, empowering customers and boosting entrepreneurs. In health care, electronic health technologies including telemedicine could leapfrog conventional delivery, especially in areas where infrastructure may be lagging, which is particularly important as populations age. In education, widespread internet and smartphone access supplies a new backbone to expand access and improve quality. With low marginal costs to scaling, technology could help even remote and vulnerable people gain access to the best teachers and curricula, allowing them to compete in knowledge and skills and supporting lifelong learning and upgrading of skills (Latin America's current poor performance in this realm otherwise leaves its citizens and enterprises ill equipped for twenty-first-century success.)

Technology can also help improve the provision of public services and urban infrastructure. "Smart cities" that feed on continuous streams of data about what is going on in the physical world can apply digital information and communication technologies to address congestion, pollution, efficiency of public services, ride sharing, and urban planning. More broadly, governments can simplify processes and improve services—in fact, our roundtable participants estimated that digitization in governance can reduce per-person interaction costs by an order of magnitude, and tackle corruption in the process. And communication technologies of course help citizens to be informed and participate in civic debate, over time leading to increased transparency and accountability.

There are also risks to these changes, however—areas in which the United States could help mitigate the negative impacts of technology disruption. For example, with digitization comes cyber intrusion, and many Latin American countries do not have experience with cybersecurity risks. This is an area where the United States has a common interest, and it can provide direct expertise and training as well as share best practices. Similarly, new technologies that permit the collection of large amounts of information about citizens by governments demand new cultural norms and systems for data privacy. This is a particularly important problem given rapid changes in governments—and even constitutions—in the region, and the keen interest that Chinese digital firms have shown in establishing a market presence here.

Social media is another space that poses enormous new challenges for the political system, as citizens participate more actively in political and social life. Roundtable participants noted that whereas a robust political process requires debate, negotiation, and consensus, today's social networks can actually make this process more difficult by encouraging the formation of groups of like-minded individuals who then listen largely to themselves, rather than developing an intermediated "common space" the way that, say, a nightly national news anchor might.

Finally there is the question of productivity and global competitiveness. Technological disruption more broadly rewards those who are most flexible, and the most able to harness and scale the tools that arise from this disruption. In a region that lags behind global averages in educational performance, this should be a wake-up call that faster development of human capital is needed here, at least to be able to understand, apply, control, and take advantage of these emerging tools.

## Technology and Demography in the "Non-governed" America

Though Latin America is on the whole more democratic and politically freer than in the recent past, many corners of the region are places where the rule of law effectively does not operate at all. Instead they are controlled by guerrillas and crime and drug-trafficking networks. The NTCA (and we include with this swath parts of southern Mexico) stands out for its weak government, poor institutions, poor infrastructure, and high levels of violence. The NTCA is particularly unprepared for the global forces that are coming, which, absent deliberate and innovative efforts, may make it even more difficult for its average citizen to generate income.

Today, two-thirds of the NTCA economy is informal, where workers earn just 35 to 90 percent of minimum wage (and generally pay no taxes). The formal economy, meanwhile, is dominated by medium and large businesses that are largely family owned, risk averse, and focused on maintaining the living standards of the owners rather than growing. Foreign direct investment suffers from weak risk-reward trade-offs, limiting opportunities for commercial development and modernization.

Human capital is at low levels. Two-thirds of students here never finish high school, and three-quarters of those who finish high school can't pass a standardized math test. Less than 11 percent go to college. And for those who do, academia is not well aligned to provide students with the skills that future industry will need.

Better policy and governance should be the obvious strategies to improving these conditions, but governments in the region are challenged by deep polarization, and they repeatedly fail to align stakeholders to support programs like housing, infrastructure, and education that are necessary to facilitate jobs in a digital economy. And with low tax revenues and collection rates, they have few resources to build these ecosystems even if the politics were to align.

Poor conditions, combined with poor institutions, have therefore increased the importance of "outsider" and nongovernmental actors to the daily lives of citizens in these least developed parts of the region. On the one hand, drug cartels operate as part-roving, part-stationary bandits bent on extracting wealth from the local population as a form of pseudo-governance until displaced by another gang. A new and far more positive phenomenon, though, is the spread of commercial, mobile, internet-based self-employment and income-generating platforms—such as the established Uber and Airbnb, but also including some locally developed mobile web–based microfranchises. These technology platforms have started to take hold in the NTCA, offering both jobs and pockets of success in improving everyday lives.

In a region where very few institutions seem to work right, these platforms demonstrate how even within a small slice of life it can be possible to create working systems. User ratings of providers and businesses, for example, offer a sense of social trust. Managed platforms minimize opportunities for the petty corruption or extortion that are otherwise common in daily business activities, and dissatisfied consumers can furthermore seek accountability for fraud through corporate, rather than (likely unresponsive) governmental, recourse. Many platforms promise functional third-party customer service, increasing a consumer's confidence to try a new product or service.

Going beyond the consumer, low-cost onboarding—and the targeted education for the platform workers themselves that goes along with it—offers novel opportunities for professional training. This is particularly valuable in a region with very low levels of human capital. Importantly, gig economy or digital sales platforms also incorporate, by default, appropriate tax payments. This sort of turnkey formalization of business and commerce helps provide sorely needed public revenues to local governments who chronically lack basic resources, but it also, over time, shifts default social expectations for the relationship between an individual and his or her government—and the mutual responsibilities that come with that. In a sense, it strengthens the social contract.

One striking example shared at our roundtable is that women in Guatemala will take an Uber ride by themselves but not a taxi ride. They appreciate the reviews and accountability of Uber but generally do not feel safe enough, or trust their government's protection enough, to take a licensed taxicab. While these tech-enabled self-employment platforms themselves cannot substitute for long-term good governance, they clearly can complement and even extend stability and infrastructure in areas where governments are able to provide a modicum of this desired good governance. These platforms are in a sense technological analogues to traditional political-economy institutions. They offer a new and compelling answer to the question of how the NTCA might jump-start its way out of continuing cycles of poverty and violence—and enable average citizens to meet the needs of other citizens more productively.

In parallel with their indispensable efforts to promote governance and security in this region, the US government and multilateral groups should therefore explore ways to support these burgeoning nongovernmental institutions. Here, Uber or Square or Airbnb or Facebook could be ambassadors of US values and strengths as important as traditional multinational mainstays like Boeing or General Electric or Coca-Cola. One could imagine how tech-enabled platforms or microfranchises could increasingly address unmet needs: not just through better bakeries, ministores, day-care centers, electrical services, plumbing services, and so forth, but also in traditionally government-mediated fields such as health care, banking, credit, insurance, education, clean-water services, and trash collection. If they can be proven among the difficult conditions of Central America, such approaches could be applied elsewhere in Latin America to improve jobs and stability.

Meanwhile, the US government should continue championing targeted investment in this region, especially to help provide the basic infrastructure for business and commerce to function. It should encourage existing—and funded—institutions such as the Inter-American Development Bank to take on loans or augment creditworthiness in these most impoverished parts of the region, rather than in the investment-grade Latin American countries that dominate their portfolios today. And where capital is put at risk in this zone of poor institutions and weak track records, it should as a rule be done in public-private partnerships to bolster accountability and quality of execution.

Finally, we have the question of violence, which has become so severe in the NTCA as to reduce the overall life expectancy of men and to act as a massive "tax" on economic activity of all kinds—and which periodically sets off large migrant outflows toward the United States. Roundtable attendees speculated whether there is an opportunity for digital technologies, alongside security-apparatus funding and traditional rule-of-law efforts, to significantly improve the effectiveness of policing and the monitoring and reporting of crime.

For example, one might imagine the application of syncretic big-data platforms, such as those developed by US tech firm Palantir and used (with some controversy) by US municipal police departments, to attempt to predict criminal activity; would such efforts, if effective, be better received in a region where the stakes are demonstrably higher? The same goes for video-monitoring facial-recognition technologies. The Chinese government has taken these tools too far in creating near police states, but they have also been widely used to reduce interpersonal crime across similarly highly urbanized areas, as in Europe, and are seeing continuously lower costs and better levels of effectiveness.

## Governance Lessons from the Emerging New World: India

Although half a world away, India's ongoing navigation of the emerging new world nonetheless offers potential lessons, positive and negative, for the diverse countries of Latin America. In many senses a country that has been inward looking in its modern history, India is now rising to major-power status, with economic and social transformations that have been truly revolutionary. India's working-age population will grow by 183 million people from 2020 to 2050. And the country has broken out of its past low-growth, poverty-driven mold to have averaged 7 to 8 percent annual GDP growth since 2004; trade and foreign direct investment flows have increased dramatically.

As Hoover colleague and former US ambassador to India David Mulford described to us, even within a single country India sees regional diversity and disparities similar to Latin America's. In an economy that remains 65 percent rural, yet is studded with sprawling megacities, per-capita income is about $2,000, expected to double to $4,300 by 2030—but in Delhi it's already $5,000, while in northern states such as Bihar it's only $600. Five of the thirty-two states account for two-thirds of all manufacturing. At its peak, the education system is world class, but here again just five states represent half of all postgraduate students, girls generally receive less schooling than their male peers, and only one-fifth of young people have any sort of vocational training. Yet, today there are still 129 million Indian women with secondary education who, given the generally low wages on offer, do not participate in paid work, leading to an overall female worker share of only about 30 percent.

So, India has clear governance challenges that echo many of those seen in Latin America. At the same time, the country's early experiences with applying technology in service of those is instructive. In response to the federal government's Smart Cities Mission, a recent opinion survey asked Indian youth what they thought should constitute a "smart city." Their response: a city that works—lights that stay on, roads that are paved, toilets that flush. This is an applied, practical vision for governance and technology, not a utopia.*

India's outsourcing sector for IT software services has grown by an average of 30 percent per year since the 1990s, but it still provides only a small share of the country's overall employment and economic productivity. A modernizing India has therefore put strong emphasis on developing a broader, more enabling technology infrastructure: national mobile telecom coverage, for example, was accomplished in three years, between 2006 and 2008. The Aadhar national ID card system launched in 2009 and now covers more than 1.3 billion people, giving easy access to the internet and a wide range of government and other services, and helping to overcome India's long-standing infrastructure weaknesses. That system underpinned a quickly announced and hastily implemented 2016 demonetization of the economy, in which 90 percent of cash currency was removed from the economy, but which was pushed through on the (likely correct) justification that switching the country to electronic payments would dramatically reduce endemic petty corruption that acts as a tax on all commerce and daily activities. Other government policies have helped establish massive new access for individuals to the internet; to banking services, including banking credit opportunities and electronic money transfer services in rural areas; and finally to an emerging nationwide e-commerce industry, including domestic and foreign firms such as Walmart and Amazon.

The Indian internet technology start-up scene appears promising, with an estimated thirty start-up firms having attained "unicorn" $1 billion valuation status. Prime Minister Modi's first administration (2014–19) accomplished many important reforms, such as stronger foreign trade and investment flows and a unified national goods and services tax regime that has brought India closer to having a single national market.

But there is significant work to do. According to Stanford Law School researcher Dinsha Mistree, India's economic sectors remain subject to protectionism, and foreign disruption often remains unwelcome: The so-called "License Raj"—a manifestation of the numerous veto points within government—still exists for technology investments and privileges established domestic players. Intellectual property protections are weak, further stifling such investment. A wave of privatization is likely needed to overcome moribund state-owned monopolies that stifle growth and investment across entire sectors, such as energy or agriculture, which have yet to share in the gains from technology envisioned elsewhere. Together these vestiges of India's previous outlook on the world have taken some of the wind out of its newer vision for economic and social modernization—something Latin American government, business, and civil society leaders with similar aims, and baggage, should watch carefully.

* Khushboo Gupta and Ralph Hall, "The Indian Perspective of Smart Cities" (2017), presented at the 2017 Smart City Symposium in Prague. doi:10.1109/SCSP.2017.7973837.

India, meanwhile, provides another model of a radical financial tech, or "fin-tech," policy—moving society toward purely digital (and therefore traceable) payments (see "Governance Lessons from the Emerging New World: India" in this chapter). The Indian public has seemingly accepted the losses in liberty associated with largely replacing cash transactions with this technological alternative as a net positive, given the country's heretofore intractable problems of corruption. Could such a strategy similarly lower street extortion in the NTCA or in other regions of Latin America facing similar governance challenges?

While crime in the NTCA and other parts of Latin America has many drivers, roundtable participants emphasized one root cause where the United States has direct influence: drugs. Discussants noted that the United States' largely supply-side-focused "war on drugs" has, despite decades of efforts, failed to stem the availability and use of illicit drugs here. This creates a black market in which Latin American criminal cartels thrive, awash in money and arms. These repeated failures would seem to call for a radical rethinking of US drug policy, including a focus on reducing the demand for drugs through selective decriminalization and aggressive treatment and education programs. Any success here over time would have hugely positive implications for the United States domestically as well as for the functioning of the most marginal of Latin American societies.

## Demographics, Migration, and Implications for the United States

Our project considered how these conditions and the future prospects for Central America and the less stable parts of Mexico might affect the pressure for emigration to the United States. Discussants observed that such pressures are affected in the long term by essentially unalterable demographic trends, but short-term variations in net migration between the United States, Mexico, and the NTCA remain dominated by changes in economics and violence.

Today, the population of Mexico is about 125 million, growing slowly (at about 1.2 percent per year), and aging rapidly. The number of children under fifteen is shrinking, and the number of seniors over sixty-five is growing: by 2050, there will be 110 seniors for every 100 children. Mexico has fallen to replacement-level fertility; its workforce will grow by 25 percent through 2035 and essentially remain flat thereafter. Improved health and declining infant mortality have already led to substantial increases in life expectancy (the average now being seventy-five years, projected to reach eighty by 2050). But as described above, homicide due to drug- and gang-related violence is a major cause of death for young men; future gains in life expectancy will depend on how these countries deal with violence.

Moving south, the combined population of the NTCA is about thirty-one million. Guatemala is growing rapidly, El Salvador slowly. Fertility-rate declines and workforce-growth trends lag those already observed in Mexico by about twenty years. Over the next two decades, for example, the Guatemalan potential workforce will explode by 55 percent and expand by another 25 percent on top of that in the two decades that follow. Fertility in the NTCA is higher in rural areas, particularly those with high shares of indigenous people, and with low education levels teenage parenthood remains stubbornly high. Were this to change, it could help increase women's participation in the labor force in the NTCA, which remains low.

What does this mean for migration? Some 12.6 million people from Mexico live in the United States today, representing nearly one-third of the US foreign-born population. Mexicans living abroad (the vast majority of them in the United States) now represent 10 percent of Mexico's population. And another three million people from the NTCA live in the United States, representing 6 percent of the population of Guatemala, 6.5 percent for Honduras, and 22 percent for El Salvador.

Net migration from both areas to the United States was very high between 1995 and 2005, due both to turmoil in Central America and labor-driven migration from Mexico. This was followed, however, by a huge decline in emigration from 2005 to 2010, plus an uptick in return migration from the United States (the result of the 2008–09 Great Recession and increased immigration enforcement during the Obama administration).

Going forward, net emigration from Mexico is expected to be well below the historical average: about fifty thousand per year to all countries of the world, including the United States, with a large component of that being formal, documented migration. In fact, Mexico today sees large numbers of returnees from the north, including children born in the United States.

Net migration from the NTCA countries, however, particularly from El Salvador, is expected to continue to be substantially greater, due in part to the climate of violence. The number of migrants from the NTCA in transit through Mexico—very few of whom end up staying in Mexico, which has a minuscule foreign-born population of just 1 percent, similar to that of Japan—is now returning to its earlier peak, reached just before the Great Recession. Most first-time migrants are between fifteen and twenty-nine years old, are more likely to come from urban areas, and have attained a higher level of education than prior generations.

These changes in flow are prompting novel policy challenges. Though the concept may sound odd to US sensibilities, the growing history and now bidirectional cross-border flows of Mexican migrants has raised the issue of a so-called shared population. The task of harnessing their social contributions and meeting their needs has largely fallen to local governments, who increasingly face questions of health-care services, education, and labor productivity. Do federal governments in either country have a strategy for effective governance of these peoples going forward?

Meanwhile, rising through-migration of NTCA asylum seekers in Mexico is changing attitudes there toward immigration as well. Once focused on US treatment of Mexican migrants, Mexican politicians and citizens themselves are now wrestling with the idea that their country could also become a destination for migrants looking for opportunity or refuge. This speaks to an opportunity for US-Mexico collaboration and coordination.

## Technological Change and a Return to Great Power Influence?

One expected theme that emerged from our roundtable conversations was the matter of growing Chinese government and business interests across Latin America. In a region that is in the United States' "neighborhood," China has made significant inroads through strategic investments and government-to-government relations, while US geopolitical attention has largely been turned elsewhere. China, for example, is already Brazil's, Chile's, and Peru's biggest trading partner. Troubled Chinese oil-for-loans deals with Venezuela are well known, but China has also made billions of dollars in high-profile policy bank infrastructure loans to Brazil, Argentina, and Ecuador, often without fiscal policy covenants attached. It is the region's largest creditor.

Changing technologies may open the door to deeper Chinese ties across the region. For example, Chinese mining firm Tianqi Lithium Corp, likely funded by Chinese state-owned banks, was recently cleared to purchase stakes in a Chilean lithium miner previously held by a Canadian firm—lithium being a crucial input of growing global importance for electric vehicles and other electronics, and traditionally considered a strategic resource for the Chilean state.

As our chapter "China in an Emerging World" explored, the country is very strong on the development and consumer application of mobile digital technologies. And Latin America is likely to become an increasingly attractive market for this. Public-security and digital-surveillance technologies, for example, are one area of growing Chinese export interest that might find receptive customers in this violence-plagued region. Advanced digital communications networks are another. What are the trade-offs of such investments, and does the United States have an interest in the choices made here?

Private Chinese internet businesses are also eager to expand in the developing world, seeing little opportunity to gain market share in the United States and Europe. Without robust homegrown alternatives in Latin America, will Chinese firms come to dominate emerging consumer sectors there such as mobile digital payments? This is an area with potentially strategic data-gathering implications (private Chinese internet firms regularly share user data with governments when compelled), and one with no clear local governmental or institutional goals; and US and European tech firms have relatively little to offer. The United States should consider how it might work with governments and commercial partners in the region, such as existing banks, on principles for the development of such mobile-payment systems—which are often linked to broader digital governance national-ID schemes, on top of which private firms can offer their own goods and services.

One of our contributors observed a historical predilection across Latin America for caudillos—political strongmen—who profess to offer magical solutions to entrenched problems. These represent a triumph of the individual personality over political institutions, aligning philosophically with the Chinese political and social tendency toward "men" over "law." Our roundtable and the paper in this volume demonstrate the numerous reasons why the United States and its democratic institutions should be interested in this region—and the ways in which those interests might be realized in a changing world. At the same time, we also observed that, even in the past, US attention toward Latin America often peaked when the United States felt that its dealings with Latin America played into broader great-power rivalries. Since the fall of the Soviet Union and the dominance of Pax Americana in defusing such concerns, US attitudes toward this region could be characterized as a sort of benign neglect, inflamed again periodically only at its own southern border. In this emerging new world, perhaps a return to great-power relations gives a new—and ultimately beneficial—reason for constructive US–Latin American engagement once more.

## Emerging, from Where?

Our Latin America roundtable at the Hoover Institution was moderated by the Honorable Pedro Aspe, who has long participated in and observed Latin American governance across a variety of administration roles, including as Mexican minister of finance, and through his involvement in a broad swath of international fora and investments. In our discussions, Aspe remarked that as we consider the longer-term transformations described here, it is also important to understand the fragile and rapidly changing Latin American political landscape from which those forces are arising. Indeed, the December 2018 roundtable coincided with some of the most relevant political events that have taken place in some of the largest economies of Latin America in decades, namely Brazil, Mexico, and Venezuela. Each of those changing foundations deserves mention here.

Starting with Brazil, the largest economy in Latin America, the recent election of Jair Bolsonaro as president signals a right-wing populist regime change with a nationalistic bent and a tendency to push for economic reform. It is important to understand Bolsonaro's positions regarding free trade and on the apparent opposition between labor protection and technification of production processes. Here, the appointment of Chicago school economist Paulo Guedes as minister of finance was a positive signal that he wants to open the Brazilian economy much as Mexico did in the 1990s, when it successfully encouraged new investment. Going forward, we should pay close attention to his economic-reform proposals and how they fit into a free-trade world. In theory, Bolsonaro's reforms can be enacted, while in practice their implementation may be more complex given the nationalistic campaign promises he made on the road to the presidency.

Then we have Mexico, the second-largest economy in the region. The election of Andrés Manuel López Obrador as president—a left-wing populist with an isolationist background—will very likely result in relevant changes to the country's economic policy. It will be of great importance to see how his proposed economic policies are implemented in a country that has experienced considerable growth in its middle class over the past two decades, most of it linked to a boom in manufacturing driven by free-trade agreements in general, and specifically by the North American Free Trade Agreement, or NAFTA. It is relevant to mention that one of the first things López Obrador did after winning the presidential election was to back the renegotiation of NAFTA into the United States–Mexico-Canada Agreement (USMCA), which its partners all eventually ratified by early 2020. He has also continued to support the Comprehensive and Progressive Agreement for Trans-Pacific Partnership, or CPTPP (a successor to the negotiated Trans-Pacific Partnership that was shelved when President Donald Trump withdrew the United States' signature in 2017), which was negotiated by and ratified under his predecessor. The CPTPP goes even further than the USMCA and the traditional concept of free trade, as it also pushes for the modernization of areas of the economy such as digital products and intellectual property, while heavily limiting the role state-owned enterprises can play. Thus, it will be interesting to see how this apparent support for free trade in general interacts with López Obrador's isolationist, entitlement-heavy, and government-centric economic policies—which echo those practiced by Mexico's Institutional Revolutionary Party (or PRI) in the 1960s. As in the case of Brazil, we have to better understand López Obrador's stand regarding the apparent opposition between labor protection and the further automation and technological orientation of production processes.

Finally, we have Venezuela, a country that continues to suffer one of the most dramatic economic deteriorations in recent history in Latin America. The start of a second term of Nicolás Maduro as president has divided Venezuela's society into the ones that consider him the legitimate president and the ones who consider his election so invalid that a different president needs to be elected. The international community, including the United States, weighed in heavily in response, mainly in favor of having new democratic elections in the country, and in refusing to recognize Maduro as the legitimate president. This may have been the opportunity for regime change in a country that thirty years ago could have been considered one of the most solid economies of the region. The precipitous drop in global oil prices in early 2020 further undermined the foundations of the Venezuelan rentier state. A regime change in Venezuela would considerably alter the country's labor and economic prospects, although any improvement is sure to take some time to be observed given the relatively high level of deterioration its economy has currently suffered.

Collectively, these developments underscore the degree to which waves of public and political sentiment can gather strength and quickly wash—or crash—through society in this region. We look to an emerging Latin America, the demographics and twenty-first-century technologies that will shape both it and its global partners, and how to best prepare and respond to that. But the experience of our moderator also reminds us that in doing so we should anticipate that the region's fragile capacity for governance will always be underpinned—or even broken down—by constantly shifting regional political realities.

For more:

- "Emerging Demographic Challenges and Persistent Trends in Mexico and the Northern Triangle of Central America" by Víctor M. García Guerrero, Silvia Giorguli-Saucedo, and Claudia Masferrer
- "Digital Transformation in Central America: Marginalization or Empowerment?" by Richard Aitkenhead and Benjamin Sywulka
- "Latin America: Opportunities and Challenges for the Governance of a Fragile Continent" by Ernesto Silva

*https://www.hoover.org/publications/governance-emerging-new-world/fall-series-issue-418*

### From "Latin America: Opportunities and Challenges for the Governance of a Fragile Continent"

*by Ernesto Silva*

Facebook Live. That was the platform chosen by Jair Bolsonaro to issue his first statement after learning of his triumph in the presidential elections in Brazil in October 2018. It was not a speech at the headquarters of his party or in a public place. It was not the television channels or the radio stations that intermediated in the communication with the citizens. More than 300,000 people saw his statement live, and within the hour more than 2 million people watched his approximately eight-minute-long message, quickly registering nearly 350,000 comments and reactions. The elected president of the continent's largest country, who has 8.7 million followers on Facebook and 2.4 million followers on Twitter, chose social networks as the platforms for his inaugural message. These were new technologies and platforms, but they nevertheless reflected a phenomenon already known: citizens supported a kind of "savior" or "solution" against their disenchantment with the ruling elites of recent years.

The fragility of institutions in Latin America continues to be a central challenge for governance. The dissatisfaction of citizens with their governments is channeled in many cases toward the search for "magical" or prompt solutions, which are usually conducted by caudillos—strongman-type leaders. This configures a scenario where the people (the leader), and not the institutions, are those who seek to respond to the problems that affect a country. The institutional environment is weak and vulnerable.

The amplification of the use of social networks and platforms in political activity has been a growing trend and has resulted in a more active and participatory citizenship in political processes. But this activation and participation has not implied—until now—relevant changes in the quality of the institutions.

## From "Digital Transformation in Central America: Marginalization or Empowerment?"

*by Richard Aitkenhead and Benjamin Sywulka*

There are tremendous systemic interdependencies that make the transition to the emerging digital and conscious economy extremely difficult. Let's begin with tax collection: the region has low tax collection (14 percent of GDP on average), of which much goes to operating the government structures (75 percent of budgets), including the salaries of hundreds of thousands of government employees (38 percent of budgets). This leaves little room for investment in infrastructure—and the little that is invested often results in poor quality given the corrupt schemes by which the contracts are awarded. The lack of road, port, and airport infrastructure in turn increases operating costs not only for businesses, but also for the health and education systems. With tight operating budgets, basic health and education services are poor both in quality and coverage. The combination of children's needing to contribute to the family's income-generation activities (almost one-fifth of the workforce is seven-to-fourteen-year-olds), poor road and school infrastructure, learning difficulties due to chronic malnutrition, and limited access to high-quality teachers results in a region where one-third of the population never finishes primary school, and two-thirds never finishes secondary school. Of the one-third that does finish secondary school, less than a quarter can pass standardized math tests, and less than 11 percent go on to pursue a university degree (only 6 percent in the northern countries of the region). This leaves most people with no choice but to work in low-skill jobs, which in the northern Central American economies (which are 73 percent informal) means earning between 35 percent and 90 percent of minimum wage. And to close the vicious cycle, none of those informal businesses pays taxes, which brings us back to the low tax collection with which we started. . . .

Using this as a starting point for strategy, the greatest unmet need for average citizens in the region is income generation. While national strategies in the region have focused on solving the income-generation problem through formal job creation, digital transformation opens up the door to solving income generation itself. As it is, in a context where two-thirds of the businesses are informal, and 80 percent of new businesses never grow beyond five employees, self-employment is already a widespread reality. What has become popular in developed economies—the "gig economy"—has been a reality for a long time in the region. In fact, it has been the only source of income for much of the population. The difference is that gig-economy platforms today are open meritocracies in which people can expand their client base and pricing through rankings and reviews. The non-platform-enabled gig economies of Central America are closed "arbitraucracies," where people are forced to succumb to pricing and market access restrictions imposed by brokers who arbitrage between supply and demand. These brokers connect marginalized people with buyers for their products and services, but their business models tend to be low-volume, high-transaction-cost models, in contrast to digitally enabled platforms that tend to use high-volume, low-transaction-cost business models. The "offline" marketplaces for products and services in the region tend to be very inefficient, friends and family based, geographically limited, nontransparent, and subject to unfair prices due to significant arbitrage. The online marketplaces are introducing transparency, trust, and efficiency into the transaction culture, and platforms like Uber and Airbnb are already generating income for thousands of people in the region.

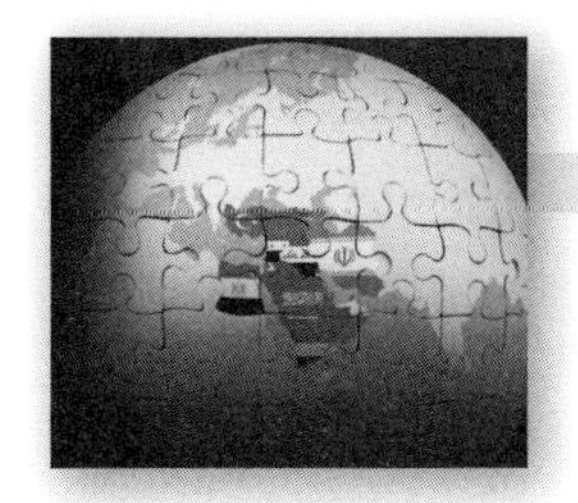

CHAPTER 4

# The Middle East in an Emerging World

When experts discuss the Middle East and North Africa, they often begin with the Shia-Sunni divide, the Israel-Palestine conflict, or the prevalence of Islamic terrorism. We know these conflicts persist. We'd like to add more dimensions. The individual countries in this region are also affected by the powerful forces that this book is concerned with: demographic change, emerging technologies, and weakening governance. Indeed, the countries we surveyed in our discussions of this region—the Kingdom of Saudi Arabia, Egypt, Turkey, the Islamic Republic of Iran, Israel, and Tunisia—are case studies of how demographics and technology can intersect to change the fundaments of a country, and the vital role governments play in determining whether that change will be for good or ill.

We will work from three general demographic baskets: the youth of Saudi Arabia and Egypt; the expiring demographic-dividend opportunities of Iran and Turkey; and the ever more diverse Israel.

## Choices to Be Made in the Kingdom of Saudi Arabia

With a young population and a technologically advanced society, the Kingdom of Saudi Arabia would appear to have a bright future. Through heavy investment in network infrastructure and next-generation technology, it can develop a thriving post-oil economy and enjoy long-term strength and stability. Or so says Crown Prince Mohammed bin Salman in his Vision 2030 plan, an expansive, multisector blueprint to build a modern Saudi economy. In his investigations for this project, Harvard political scientist (and a member of the Moroccan royal family himself) Hicham Alaoui took a less Pollyannaish view of the future.

**FIGURE 10 Egypt's youth bulge brings uncertainty.** Egypt may face continued volatility stemming from a population that remains relatively young, with half of Egyptians under age twenty-five. But Turkey and Iran, the region's other most populous countries, are rapidly aging.

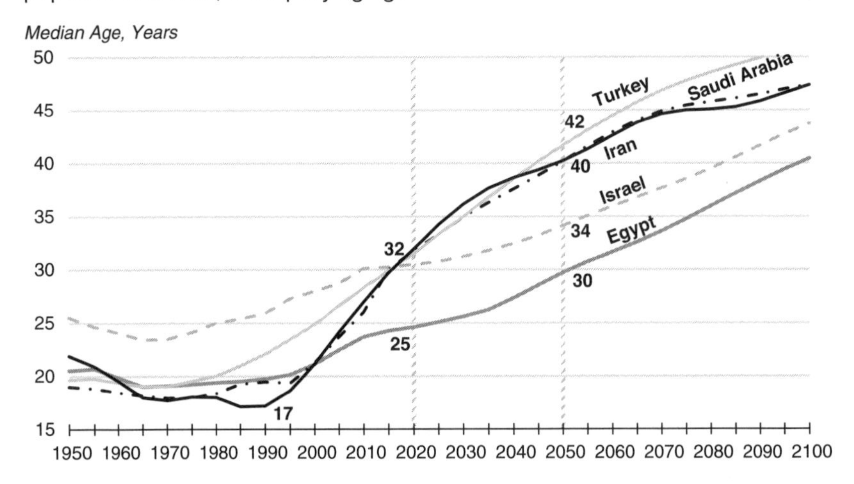


**FIGURE 11 Opportunities are missed in the Middle East.** The rising number of workers per dependent offered a potential economic boost in Turkey, Iran, and Saudi Arabia, but these countries were not able to capitalize on the favorable demographic trends, which are now diminishing.

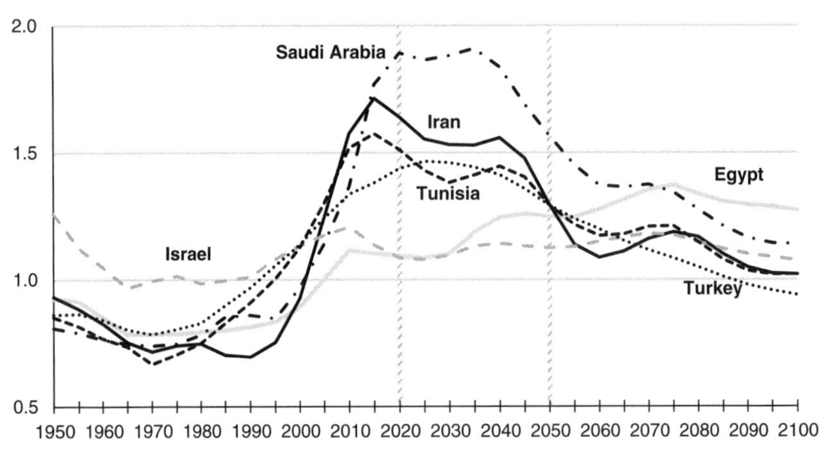


Data Source: United Nations, *World Population Prospects: The 2019 Revision*, medium variant. Demographic analysis by Adele Hayutin, 2020.

The crown prince is correct about the demography and network infrastructure of Saudi Arabia. Saudi Arabia is young. A moderate (for the region) official median age of thirty includes a substantial share of migrant guest workers; among Saudi nationals alone, however, roughly 60 percent are under the age of thirty, and 40 percent are under eighteen. These youth, centered in urban areas, tend to lean more liberal than members of older cohorts. They supported Bin Salman's decision to allow women to drive, and the majority want more rights for women. And the younger populations tend, as is often the case, to be tech savvy. However, they have limited economic opportunities and express dissatisfaction with the state of the kingdom. At least a quarter of them are unemployed, and many find themselves stuck between school and a job, without having started a family, in what Alaoui called "waithood." Those who do have jobs tend to find them in the public sector, which employs two-thirds of Saudi nationals, or adjacent industries. Saudi workforce participation by women, once lagging, is growing and has now exceeded the regional average over the past decade. This will contribute to economic growth. Yet female workforce participation remains low in an absolute sense—just two-fifths of the United States' rate, for example. Saudi youths are ambivalent about the future: just 51 percent think they will enjoy a higher standard of living than their parents.

The Saudi population enjoys good broadband availability, mobile penetration, and network connectivity. There are 122 mobile phone subscribers for every one hundred people, and, with over 90 percent of the population using the internet, three-quarters of Saudis are active on social media. As we have seen around the world, access to mobile and internet technology has yielded mixed results. Saudi society can escape some of the regime's traditional censorship by turning to social networks—women in particular can express themselves online in ways they have not been able to offline—but the regime has asserted its authority over the digital space. It has long regulated media and television access and enforced speech restrictions, and Bin Salman has introduced extensive cyber policing. The regime now monitors internet activity and has imprisoned activists for online speech. More broadly, the regime views access to the internet or mobile connectivity as it does political and civil rights: privileges to be extended, not rights granted to all.

So perhaps Saudi Arabia's future does not necessarily appear all that bright. In contrast to the optimistic outlook of the crown prince and his Vision 2030 plan, Alaoui warned that political instability is just as likely. Bin

Salman hopes to move Saudi Arabia beyond its existing rentier system, where about half of GDP is now based on the sale of oil. He recognizes that hydrocarbon revenues will decrease, and the royal elites may not be able to rely much longer on the provision of oil rents in return for political stability. So he looks to dynamism from the country's favorable ratio of workers per dependent—a product of youth, but also of a large migrant labor population—and technology to transition the kingdom into a newly productive era.

However, in Alaoui's words, Bin Salman "desires modernization without modernity." He wants the trappings of a contemporary, technologically advanced society but is unwilling to surrender control over the direction of the country or to democratize. It is not clear that these two desires are compatible, and the direction of Vision 2030 so far would suggest they are not. State-driven investments in technology remain preeminent, most notably in the crown prince's vision of a $500 billion "high-tech" city in the desert. And the economy remains oriented around the largesse of the state and, in particular, royal elites. Alaoui describes the relationship thus: "The state will give technology on the basis that it decides; and you will consume it how the state says you can."

It is possible that as today's young generation ages, it could force greater political openings and economic modernization. But while tech savvy and socially liberal, this group is not yet particularly politically active, nor do they seem eager for economic liberalization. On the contrary, some two-thirds of Saudi youth support an expansionary monetary policy and seek more government largesse. Given the lack of economic opportunities, it is easy to understand why they might shift toward statist solutions in this way.

Until recently, the Saudi monarchy was widely accepted as the legitimate sovereign in Saudi Arabia, but now it rules as a police state. If the youth cohort, now aging, were to trigger political dissension or uprising, what might that look like? Though the young may spark political uprisings in the Middle East and North Africa, they rarely direct the outcomes. The youth were out in the streets of Tehran in 1979, but Ayatollah Khomeini was not the handpicked choice of the student protesters. Large populations of young people, armed with technology and unhappy with the status quo, breed uncertainty, and with that in mind, Alaoui predicted that "a future of instability and volatility is as plausible as the prosperous and secure outcome evoked by the Saudi leadership today."

A comment on Saudi Arabia's foreign policy and its role in the region is appropriate here. The crown prince has adopted an assertive stance and increased his country's engagement in regional affairs. Saudi Arabia is balancing against Iranian influence around the Middle East, particularly in Iraq; has been intervening in the Yemeni Civil War for years now; and has moved from cooperation to confrontation with Qatar. Roundtable participants noted that Bin Salman views foreign policy through the same lens as domestic policy: he wants Saudi Arabia, and by extension himself, as the primary authority in the region. Saudi efforts to keep Iran in check and provide support for friends in Yemen and Tunisia reflect that ambition. But Saudi engagement, much like Iranian engagement, has tended to disrupt stability, and will continue to do so. Given his relative lack of success, and following the murder of journalist Jamal Khashoggi in October 2018, the crown prince has lowered his profile abroad. Yet he continues to seek legitimacy and authority, expanding ties with Russia and China to improve his own position.

Many of the themes of this book are met in Saudi Arabia. It is navigating a demographic transition, and its leadership has staked the future of the country on emerging technologies. There is great opportunity in the kingdom, but also great uncertainty and risk of political instability. What may define its future is governance, specifically the policy choices made by the crown prince and the House of Saud. Will they recognize that a thriving economy and technological dynamism are incompatible with a state-centric, rent-based economic system and a social system with deep gender, religious, and country-of-origin biases? Will they seek to copy China's model of entrepreneurial growth alongside strict state censorship and state-based internet sovereignty? Or does perhaps the most likely outcome rest somewhere in the middle, with the House of Saud digging into its reserves and attempting to maintain the status quo—and the existing state-society relationship—as long as possible?

## Staying Ahead of the Population in Egypt?

What was said of Saudi Arabia was echoed during our discussions of Egypt, but in more dire tones. Egypt faces similar economic and demographic conditions to the kingdom, though they are arguably more severe and leave less room for optimism. But, as in Saudi Arabia, the challenge for Egypt ultimately comes down to the policies and characteristics of the government. The country has a large, young, and rapidly growing population (see Figure 10), but the regime of President Abdel-Fattah al-Sisi seems more interested in controlling that population than helping it flourish.

In many respects, Egypt today also resembles Egypt at the time of the Arab Spring, when a series of protests and uprisings fanned across the Arab world in the spring of 2011, leading to the overthrow of Egyptian president and strongman Hosni Mubarak. As our Stanford University colleague Lisa Blaydes observed, Egypt has experienced immense political disruption in the intervening years, but there has also been a good deal of continuity. The demographic and economic conditions that predated the 2011 uprising remain, and in some senses have gotten worse. There remains a large youth population—here too, almost 60 percent of the society is under the age of thirty—but economic opportunities are few. And the political situation has also gotten more dire. Al-Sisi's government is more oppressive than ousted president Mubarak's was. He has, as Professor Blaydes put it, "institutionalized authoritarianism": established an autocracy supported by the military, passed laws restricting nongovernmental organizations (NGOs) as well as protests and political demonstrations, and enacted a harsher penal code. To go even further, the regime hobbles personal freedoms more than before and engages in mass surveillance and media censorship—it pushed through a law in 2018, for example, that allowed government surveillance of thousands of social media accounts. At the same time, the country's legislature has grown weaker, allowing for greater centralization of power in the head of state.

The general picture of Egypt is of a predatory, military autocracy that wants to exert control over the country's path. It views its youth as a national-security threat and wishes to keep it under control. But despite al-Sisi's intentions, Professor Blaydes argued, forces outside his control—demographic changes, economic opportunity, public health, and climate change—will drive the future.

Egypt is in the same situation as a number of countries surveyed over the course of this project—especially its neighbors in North Africa: it has a young and rapidly growing population but offers minimal economic opportunities. The median age is just twenty-five, and Egypt's population is set to explode by over 50 percent by midcentury. Fertility had declined in the lead-up to the Arab Spring but increased in the wake of it, in part because the Muslim Brotherhood de-emphasized existing family-planning policies during their post-Mubarak stint in power, and in part because al-Sisi's anti-NGO initiatives restricted access to and the supply of birth control.

However, even as they grow in number, Egyptian youths do not enjoy good economic prospects. Lisa Blaydes cites the fascinating results of the Survey of Young People in Egypt, conducted by Cairo's Population Council. The survey reported low rates of employment among those aged thirteen to thirty-five. Of those not in school, only 40 percent had reported working recently. And of those employed, 70 percent said their jobs require no real skills, while only 10 percent reported that their jobs required a computer. The survey results paint a general picture of dissatisfaction with the labor market, low productivity for those who are employed, and minimal confidence in the country's direction. Where Saudi youths were torn on the country's outlook, Egyptians are convinced theirs is negative: only 30 percent of them think the country's policies will leave them better off than their parents.

So we see in Egypt a large, growing, and tech-savvy population of young people, but a weak labor market and real anxiety about the future. And there is little evidence the al-Sisi government has the intention or, at present, the capacity to improve the situation. A general concern is that the al-Sisi government lacks both the resources and the will to keep the nation's infrastructure up-to-date for such a large and rapidly growing population. In some ways, al-Sisi's government seems more inclined to remove itself from society than to respond to society's needs—as seen in its decision to move the administrative capital thirty miles into the desert, away from the population center of Cairo.

The governance challenges for the al-Sisi regime go beyond demographic and economic conditions. Egyptian society is dealing with an ongoing public-health crisis. Egypt has among the highest rates of obesity and hepatitis C in the world: 35 percent of adults—some nineteen million people—are obese and a similar number are classified as overweight, which means Egyptians also exhibit high rates of obesity-related diseases, including hypertension, diabetes, and heart conditions; and 10 percent of the population is infected with hepatitis C. Fortunately, this is one area in which the government has shown initiative and progress. For example, it has implemented successful programs to screen for and treat hepatitis C, but the crisis consumes one-third of the national health budget, restricting resources available for other needs.

These social issues could be further magnified through the deleterious effects of climate change, which presents an overarching challenge. Egypt, almost entirely dependent on the Nile for water, could reach water scarcity as soon as 2030. On top of that, Ethiopia's plans for a dam on the Nile further threaten the country's water supply. Even a short-term undersupply could have significant effects on the country's stability. As Blaydes explained, the country has historically been highly sensitive to disruptions to its agricultural industry, and political or social unrest has followed food shortages or price spikes. Adapting to a changing climate—the need for new agricultural infrastructure, flood controls, air conditioning, and infectious-disease response—will act like a new tax on everyday Egyptians, and any coordinated response will only further strain the country's public budget.

Al-Sisi may hope to direct the country's future, but to do so effectively he will have to address his country's demographic and economic challenges. Given the scale of population growth and existing financial problems, discussants at the roundtable were concerned that the country simply could not generate sufficient economic growth to alleviate these pressures. And they worried what that might mean for the country's neighbors, especially those across the Mediterranean. Already, many young people now emigrate—primarily to Europe. As migration pressures grow, it is a good bet that those numbers will rise. Europe, already feeling the consequences of Egypt's difficulties, will have to prepare accordingly.

In this way, the broader themes of the emerging new world appear again. Similar to that of Saudi Arabia, Egypt's current situation cries out for good governance. Continued improvements to the country's public-health initiatives, reversals of al-Sisi's restrictive and predatory policies, and a dedication to reducing corruption could all improve prospects for economic growth while alleviating some of the severe social pressures at play. And Egypt's struggles radiate outward; whether we like it or not, they will be felt in Europe and in other advanced democracies.

While Saudi Arabia and Egypt have large cohorts of young people—and the attendant opportunities and challenges—Turkey and the Islamic Republic of Iran are at the other end of the spectrum. Both populations are aging rapidly, and both countries seem bound to let the opportunity of a demographic dividend pass by unseized.

## A Governance Deficit in Turkey

Let us look next at Turkey. With a median age of thirty-two, Turkey is young relative to its neighbors in Europe but relatively old for the Middle East. It is currently at the peak of its demographic window, as expressed by the share of its population that is working age, and those favorable demographics will diminish through the 2020s as that share declines (Figure 11). In our discussions, Aykan Erdemir, senior director of the Turkey Program at the Foundation for Defense of Democracies and a former member of the Turkish Parliament, argued that Turkey suffers from a governance deficit, which stems from its hypercentralized political system. And he explored what the country's demographic transition, coupled with the technological advances of the twenty-first century, might mean for the country's future, both internally and as a member of NATO.

According to Dr. Erdemir, the roots of Turkey's hypercentralized politics lie in the post–Ottoman Empire transition. Following the collapse of the empire, Turkey centralized authority in the state, fearful of delegating power outside the capital or of accommodating ethnic minorities. The state took on a heavy-handed and corrupt role in markets, which has impeded economic development. Under President Recep Tayyip Erdoğan, the state's centrality has only increased, and the stabilizing institutions have grown weaker. There are few checks and balances on presidential authority; one of the few, the military—once a primary constraint on the power of Ankara—has been gutted, and two-thirds of the officer corps purged. With the military weakened, Islamists at home and among the Turkish diaspora have accrued more authority, pushing the government more toward a revolutionary theology.

However, Erdemir noted in the roundtable discussion, the coercive capacity of the state should not be misunderstood as a sign of strength. By contrast, he described Turkey as having "a strong society with a weak state." Erdoğan has increased the state's brutality but also its governance deficit.

At the same time, the state is undergoing two transformations: demographic and technological. To the former: as described above, Turkey is gradually transitioning from a young to an aged society, with an absolute decline in the number of children, and is, alongside that, moving from a sending country (with an outflow of guest workers) to a receiving one (which receives refugees). As we have seen with its regional neighbors, Turkey has not given its youth the tools to succeed. Education prioritizes religious-national training over the skills necessary for success in the twenty-first century, and 33 percent of young adults are neither employed nor in education or training. Moreover, with falling fertility rates and limited opportunities among native Turks but growing numbers of refugees and Kurds, the country has witnessed a wave of anti-immigrant and anti-Kurd sentiment.

On the technology front, Turkey is a relatively connected and advanced society, but the government works hard to rein in media and internet platforms. The population widely uses social media, but Erdoğan controls roughly 90 percent of print and visual media, and uses social media to manufacture consent. The government collects individuals' data and imposes one of the world's strictest digital environments (neither Wikipedia nor PayPal is available there, for example). Erdemir described Erdoğan's approach as reflexive control of the internet.

However, the government's ambitions outpace its capacity. Erdoğan may seek reflexive control, but digital technologies have nonetheless given the population more avenues for political activism and mobilization. That could be seen in the 2019 elections, when, despite the president's best efforts, the opposition party won large gains, even defeating Erdoğan's handpicked mayor of Istanbul.

Turkey has a history of top-down redesigns of the political order, and it is going through a similar process today. The government is dismantling democratic institutions, regulatory agencies, the rule of law, education, and financial regulations. But it has not stamped out democracy—again, the last round of elections shows that opposition is alive and well. The question for Turkey is whether the demographic and technological changes under way will deepen the problem or put Turkey back on track to better governance. During our roundtable on the topic, discussants with intimate knowledge of the political dynamics in-country gave reason for optimism. They reported that although Erdoğan has worked to undermine existing institutions and bulwarks of opposition, reformers, especially secularists, have kept the memory and principles of those institutions alive. They have been able to organize online and resist some of the kleptocratic behavior in Ankara.

The West, especially Turkey's NATO allies, will have an important role to play in supporting the future of democracy and prosperity in Turkey. Russia has sought to use Turkey to foment discord within NATO—making it, in the words of one roundtable participant, "their man in NATO." But NATO, and especially the United States, would be wise to resist that effort. Democratic restoration may be a fundamentally domestic process, but European and Western powers can help build the conditions to allow that process to flourish. In the near term, we could wield the carrot and the stick of NATO: offer alternatives to Russian military systems, namely the S-400 antiaircraft radar and missile battery, while making sure Erdoğan understands the costs of breaking alliance practice. In the long term, recall the role the United States played in establishing the Turkish democracy in 1946. The United States' guidance and support helped build the foundations of democracy there. Could it help reestablish them today?

## Missed Opportunities in the Islamic Republic of Iran

If the House of Saud wants to know what can happen when an authoritarian regime lets a demographic dividend opportunity pass by, it can look to its longtime rival, Iran. Abbas Milani, our Hoover Institution colleague and codirector of the Iran Democracy Project here, and Roya Pakzad, a researcher in Stanford University's Global Digital Policy Incubator, outlined the sad state of affairs in the Islamic Republic: an aged, autocratic regime faces serious economic, political, and legitimacy crises.

Iran has experienced one of the most dramatic reductions in fertility rates in recorded history. According to one Iranian census estimate, in the ten-year period from 1986 to 1996, its total fertility rate plummeted from 6.2 to 2.5 births per woman, the result of aggressive state family-planning programs alongside urbanization and changing Iranian social mores around female education and delaying the age of marriage for women. That fertility rate is now around 2.0. In the early 1980s, Iran's population grew at an annual rate of nearly 4 percent; it is now down to 1.2 percent. This deceleration of population growth is an echo of the rapid fertility decline. This fall has been so steep that Iran actually now joins many mature economies around the world in wanting to reverse the trend. Despite the regime's best efforts over the past decade, fertility remains stubbornly low.

A result of those once-high fertility rates was that the country's working-age population grew by fourteen million people from 1995 to 2010. It continues to grow, but growth is slowing. Moreover, the working-age population is graying: men and women aged forty to sixty-four are expected to outnumber those age fifteen to thirty-nine by 2028—and that aging process will pick up speed come 2040. From then until 2050, the ranks of those sixty-five and older will swell in number while those of working age will shrink.

One might expect Iran to be enjoying favorable economic conditions given the rosy demographics of recent history, but instead it has seen high rates of unemployment and low rates of workforce participation. Iranian official data show that only 40.3 percent of working-age Iranians—or 26.6 million people—are actually in the labor force. Meanwhile, over a quarter of the young workforce (defined by those aged fifteen to twenty-nine) is unemployed. There is a significant gender disparity as well, with a dismal female participation rate of only 16 percent compared to 64.5 percent for men.

The regime has encouraged the expansion of secondary education to offset unemployment of young adults, but it has wasted much of the return on that investment. Half of university students are now women, but female university graduates have a 60 percent unemployment rate. One-third of university students are in engineering, with an even larger number of science, technology, engineering, and math (STEM) students, but Iran experiences brain drain, due in large part to the repressive regime and to the country's limited work opportunities.

All told, Iran has endured forty years of double-digit unemployment and inflation. It is dealing with a water crisis, which has triggered food insecurity. Poor infrastructure contributed to significant flooding in recent years, the effects of which were compounded by limited and ineffectual government responses. Low oil prices, coupled with sanctions and mismanagement, have limited the value of Iran's hydrocarbon resource base. And the financial system appears to be collapsing. Financial institutions charge 25 to 30 percent interest on loans, but the loans go to clerics and military officials, making them unrecoverable. Savings and loan institutions are wracked with debt, and the government has assumed much of it.

Dissatisfaction with the status quo—unpaid wages, rising inflation, natural disasters, and limited employment opportunities—triggered mass labor strikes and protests in 2018 and again in 2019. Protesters took to the streets in the latter half of 2019 and were met with violent government crackdowns. A reckoning might appear imminent. But our discussants warned that Iran is not on the verge of collapse, nor is the regime as pressed as might appear.

Consider digital technologies. Iran has a robust digital infrastructure, with high rates of internet penetration and mobile access, as well as social media use. Men and women alike make use of social media platforms and the internet. At the same time, the regime maintains effective cybersovereignty. It traffics in surveillance and censorship of online content. In general, it continues to invest in digital infrastructure and strengthen the internet, communications, and technology sector but also aims to consolidate its influence over that sector. Roya Pakzad sources this goal to policies formed in the mid-2000s under then president Mahmoud Ahmadinejad. Following the Green Movement, which arose in 2009, the Iranian Revolutionary Guard Corps (IRGC) expanded its control over domestic networks and its own cyber and information warfare capabilities. It has since engaged in such operations against targets in the United States and other Western countries, and within the region, as well as against domestic opposition.

Iranians have always resisted surveillance and police-state action. As many as 60 percent of them now use virtual private networks and similar techniques to evade firewalls and censorship, and to communicate with the diaspora, but the regime maintains its position. The people's commitment to resisting, or at least evading, the revolutionary regime's surveillance raises questions about how long the IRGC and ayatollah can maintain the cybersovereignty model. Despite their investments and best efforts, they still have relatively unsophisticated cyber capabilities compared to, for example, those of the Chinese Communist Party.

In Iran, as in Saudi Arabia, the regime aims for modernization without modernity. It seeks extensive cyber and information warfare capabilities, and it wishes to develop a powerful domestic technology sector. But it is unwilling to allow that sector to develop organically. Iran's commitment to its nuclear weapons program has invited sanctions and severely limited the country's potential (through lack of global trade, investment, and other foreign exchanges), and it continues to funnel resources to military and domestic security services and foreign-policy goals—exemplified by its support of Hezbollah and the Assad regime in Syria—at the cost of domestic well-being. Yet its challenges may just be beginning. The population is aging, and the working-age population will decline. These are real problems for Iran, but the Iranian leadership does not seem ready or willing to deal with them. How far can a revolutionary, autocratic regime go when it faces these kinds of headwinds?

## Governance Hardware versus Social Software in Israel

We have so far discussed autocratic regimes with significant governance deficits, so let's turn our attention to an exception in the region. Israel is, in many ways, a model country. It has a thriving high-technology sector and a strong economy. It is a healthy democracy, with free and fair elections. But these are, in the words of our Hoover colleague Arye Carmon, the "hardware" of a democracy. What Israel needs more of is the "software" that sustains a democratic society and enables it to effectively govern over diversity.

Israel's hardware achievements are substantial and notable—particularly in light of its unforgiving neighborhood. It has built the material foundations of government agencies, ensures open elections, and maintains a strong military. One-tenth of the workforce is in the high-tech industry, and technology accounts for 35 percent of its total exports. Israel has more than one hundred companies on NASDAQ, and technology companies from around the world come to the country to set up research-and-development centers—and to take advantage of Israel's exceptional human talent, where, for example, the female labor-force-participation rate is three times the regional average. And its surprisingly healthy fertility rate, the highest in the Organisation for Economic Co-operation and Development (OECD) member countries, promises a balanced path of gradual population and workforce growth to continue to fill those jobs. Underlining Israel's advances in technology are a healthy business environment and a culture of innovation. Israel's tech sector grew organically out of the industry built around military and intelligence development after 1973. Israel's advanced water industry allows it to live in an arid environment, and the country has advanced space, AI, and additive-manufacturing companies.

Arye Carmon worries, however, that the country's exceptional hardware may not suffice in the absence of more-substantial software. Israel is, in his words, like a ship of state that left the shipyard early. It is a young country, populated by a people that, having spent centuries in exile, are relatively new to political sovereignty. Its population has grown tenfold since its founding, to 8.5 million, and is expected to increase by half again by 2050. This populace is a kaleidoscope of unharmonious identities. Three-quarters of Israeli citizens are Jews and 20 percent are Arabs. But religious orientation is really the key identifier: 55 percent of Israeli Jews are secular, 26 percent are traditional,

10 percent are Orthodox, and 9 percent are ultra-Orthodox. Israeli society is a diverse mosaic of heterogenous minorities, but that mosaic is showing fractures. Since the founding of the state, the concept of "Jewish and democratic" has been a defining part of Israeli identity. Sadly, a religious counter-revolution has complicated the former element, and, as in many democracies, populist nationalism challenges the state's democratic values. Rapid population growth in the Israeli-occupied territories, which are predominantly Arab areas, poses further potential long-term challenges to Israel's democracy.

The lack of national democratic software makes it harder to hold the ship of state together against those tensions. The shared Hebrew language helps establish a national consciousness, but Israel has little beyond that to bind it. It lacks both a constitution and a bill of rights. Its territorial identity remains in flux—without an agreed eastern border, there is a transience to the state. It does not have all the conceptual fabric that can hold a country together and stabilize governance across such a diverse society.

The structure of the government does not help matters either. Israel's is a parliamentary system, but it is unicameral and composed of just one district. Over time, Israel's two major parties have declined, and the discord around the frequent Israeli elections of the late 2010s reflects the breakdown of politics into a host of smaller and more vocal parties and political coalitions. In the absence of a few major defining parties, Israeli politics, again resembling many other democracies, has become personalized: as Arye Carmon wrote in the pages of the *New York Times*, "Israelis no longer vote for party platforms; they vote for personas." In the Knesset—Israel's parliament—the governing coalition has lost the ability to get bills through. From 2002 to 2012, few proposed laws passed, and those that did tended to be put forward by individuals, not parties or governments. Social media has no doubt contributed to this process, as in other states, by exaggerating the fractures in political discourse, but the underlying social fabric and weaknesses of Israel's polity have contributed as well.

The challenge for Israel will be one faced by governments everywhere in the emerging new world: how to govern over diversity. Israel's national identity is in flux, with democratic values weakening and the role of religion in society being pushed to the extremes. But it is still a young state with an ancient people. That truism can be both cause for concern and a reason for optimism. Israeli society grows more diverse, but it has a strong foundation upon which to build a stronger social fabric and the democratic software it needs.

## Human Potential across the Region

The Middle East and North Africa have long been at the center of geopolitics, and the direction of the region has often followed its three most populous states: Iran, Turkey, and Egypt. So the situations in those states, described above and in this volume, are concerning. But there are positive forces at work in the region.

In our discussions for this project, we were fortunate to have the participation of a remarkable young Tunisian named Houssem Aoudi. Aoudi spoke from firsthand experience on the state of innovation and entrepreneurialism in the region. His home state of Tunisia has led North Africa in developing the civil society and democratic foundations that allow for real innovation to take hold. It has adopted some legal protections for private-sector entrepreneurs and now supports public-private partnerships. But Tunisia is not alone. In Israel, the survival instinct takes on a more national character, but its desire to remain strong and independent in response to existential threats has helped make it the world leader it is today in risk-taking inventiveness. In Egypt, the government has simplified the process of starting new businesses and supports entrepreneurship; there are now incubators and angel investors funding start-ups there. Aside from Israel, the regional leader in innovation is the United Arab Emirates, in particular Dubai. Dubai has a comparably excellent business environment and strong capital flows, enabling the kind of large-scale innovation that does not occur in Tunisia or other weaker states.

Of course while achieving high-tech innovation at the frontier of human science and engineering is rare in many countries, the entrepreneurial spirit nonetheless exists everywhere, and it often takes the form of "survival innovation"—something we described in earlier investigations of Central America. In Tunisia, for example, the informal economy dominates, so people seek ways to smooth financial transactions through payment apps, among other lines of effort. Such innovation may not grab headlines or be transformative, but it is a sign of the potential of technology. When given minimal improvements to the rule of law, protection of property rights, and civil society, entrepreneurs of all stripes leap into action.

Perhaps most promising for the region, in Tunisia as with many of its neighbor states, there is an ocean of largely untapped human potential: women.

Throughout the roundtable, presenters and participants noted the large numbers of young women in the region who are slowly gaining access to education and to economic freedoms. Supporting them and giving them the opportunities to contribute to their own societies is likely the most indelible change that could happen across the Middle East and North Africa.

The challenges facing many countries in the region are problems of their own making. They are the products of poor governance and bad policies, those that prioritize authoritarian control over practices that could foster economic growth and democracy. The countries surveyed above have the potential to grow stronger and healthier. By reducing predatory government practices, ensuring the rule of law, and broadening women's participation in society, governments could improve their countries' outlooks and strengthen regional stability. Better governance will not salve all ailments—the Shia-Sunni conflict will not be resolved through better domestic policies alone—nor will it come easily. But it is, as one of us often says, using a phrase we will come back to, a "work-at" problem. Incremental improvements to government policies add up, and together they can do wonders for a society and a region.

For more:

- "Youth, Technology, and Political Change in Saudi Arabia" by Hicham Alaoui
- "Innovation and Entrepreneurialism in the Middle East and North Africa: The Cases of Egypt, Tunisia, and the UAE" by Houssem Aoudi
- "Challenges to Stability in Egypt" by Lisa Blaydes
- "Building Democracy on Sand: The State of Israel in the Twenty-First Century" by Arye Carmon
- "The Impact of Demographic and Digital Transformations on Turkey's Governance Deficit" by Aykan Erdemir
- "The Islamic Republic of Iran in an Age of Global Transitions: Challenges for a Theocratic Iran" by Abbas Milani and Roya Pakzad

*https://www.hoover.org/publications/governance-emerging-new-world/spring-series-issue-519*

### From "The Islamic Republic of Iran in an Age of Global Transitions: Challenges for a Theocratic Iran"

*by Abbas Milani and Roya Pakzad*

To accommodate the youth bulge, the Iranian regime has largely expanded the number of universities and colleges around the country. As a result, in a mere three decades since 1990, the capacity of universities in Iran has increased from less than half a million to about 4.5 million. Because of this rapid expansion, upwards of 10 percent of the adult population already has a bachelor's degree or higher. It is expected that by the middle of the next decade more than half the young population (aged twenty-five to thirty-four, male and female) will hold a minimum of a bachelor's degree. Pursuing education beyond a bachelor's degree, as a move to defer entering the inhospitable labor market, has become more common in Iran. Today, almost 40 percent of the students in bachelor's programs will continue their education toward more advanced degrees. As the peak of the population is moving higher in the age pyramid, and as the number of universities in the country has mushroomed, competition for students in many of the smaller colleges and universities has already started.

The presence of women in these university programs has arguably seen the most remarkable increase. Today, women constitute about half of university students in Iran. With about a third of university students enrolled in engineering, the number of graduates in STEM subjects (science, technology, engineering, and mathematics) in Iran is ranked fifth in the world after China, India, the United States, and Russia (all of which have a much larger population than Iran).

However, having a highly educated young labor force does not automatically translate to economic growth. Among other factors, it is contingent on bringing an end to crony appointments and quotas for members of the IRGC, the militia-cum-gangs called the Basij, and other members of the political elite. Just as the tight job market has led to the rapid rise in the average years of schooling for youth, it has also led to the unfortunate fact of Iran's leading the world in brain drain over much of the last decade.

CHAPTER 5

# Europe in an Emerging World

After sparking two world wars that brought horrific destruction to its own ancient civilizations, Europe finished the twentieth century riding a wave of economic and political success. With decisive economic, political, and military support from the United States, the fifteen countries that would form the European Union (EU) had rebuilt themselves and helped the United States prevail in the Cold War. They gradually would welcome thirteen more countries into their organization, which became widely seen as a pathway to prosperity and a guardrail against the embittered, competing nationalisms that had led to war. Some members of the EU even adopted a common currency, in part to emphasize the benefits of nations working together through economic cooperation rather than trying to dominate one another.

Today, Europe continues to exercise significant power in global trade and international politics. But the sweeping technological change of the last two decades—change that is led, even within Europe, by outsiders—has shaken European confidence. The technological revolution's threat to established employment patterns and the social order joins threatening demographic trends and volatile immigration flows to turn reborn nationalism into a new, destabilizing force in European politics (see Figure 12). These fresh vulnerabilities also deepen concerns created by the globe's changing climate and the decline of US-Russian cooperation in limiting nuclear weapons.

While much has changed in this emerging new world, this reality remains: Europe continues to be the meeting ground of the world's most important power rivalries, be they geographic, technological, or ideological. The nature of those rivalries is shifting, and expanding, in an increasingly unstable fashion. The European Union's divided system of governance—in some ways an asset during the Cold War—is coming under severe strain in this new era.

**FIGURE 12 An aging Europe confronts labor and productivity questions.** Europe will remain the world's oldest region. The burden of aging populations and shrinking workforces will increase calls for advanced technologies, delayed retirement, and additional migrant labor.

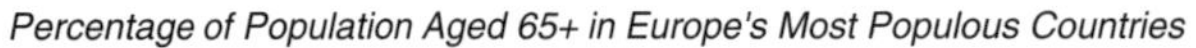

*Percentage of Population Aged 65+ in Europe's Most Populous Countries*

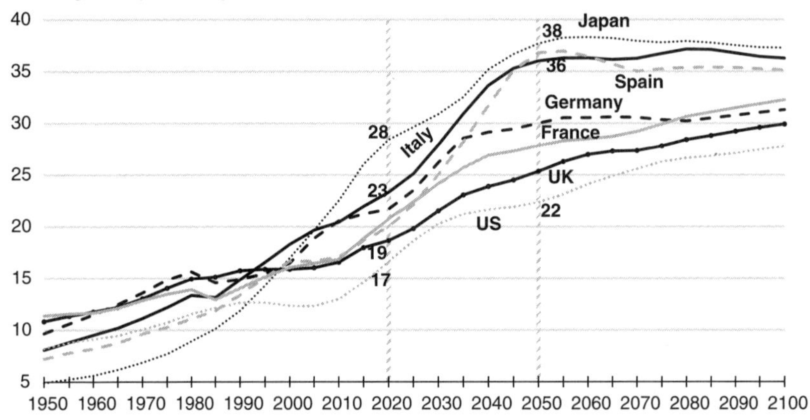


**FIGURE 13 New challenges for European immigration.** Germany, Italy, and the United Kingdom saw large influxes in migrants during the decade 2010 to 2020, with a combined flow of 1 million per year. Migration from new sources will require better integration policies across Europe.

*Net Number of Migrants during Five-Year Period Ending in Specified Year, in Thousands*

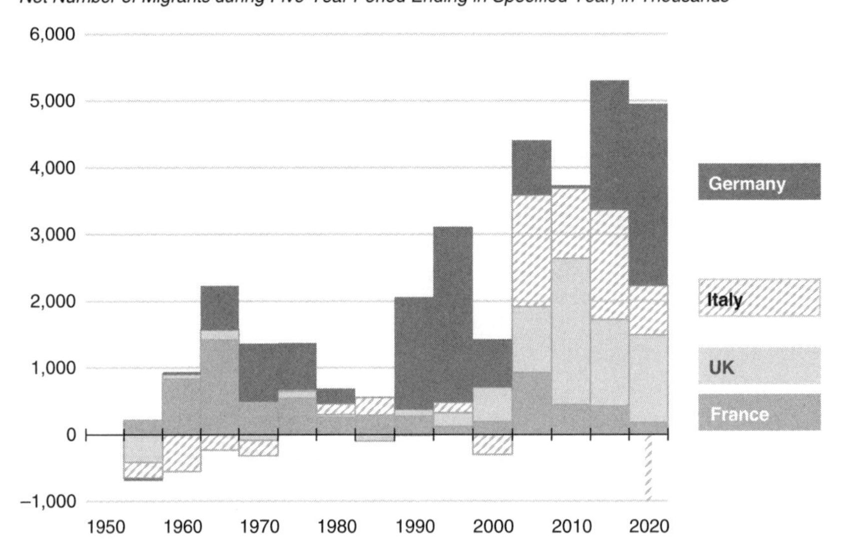


---

Data Source: United Nations, *World Population Prospects: The 2019 Revision*, medium variant. Demographic analysis by Adele Hayutin, 2020.

The number of American and Russian tanks and missiles deployed along European frontiers was once judged as the decisive marker of geopolitical power. The last act of the Cold War turned on the American decision to deploy Pershing missiles in Europe to counter Soviet SS-20 intermediate-range rockets. Today, US and Chinese corporations compete to gain control of technological assets and dominate markets while forging new political alignments in Europe, where Chinese investment in and loans to NATO members—such as Italy and Greece—have helped create political support for Beijing's "One Belt, One Road" strategic initiative. Russia, meanwhile, with relatively inexpensive outlay, has weaponized disinformation via digital platforms on the continent to weaken the resolve and cohesion of NATO, while investing in updates to its conventional and nuclear arsenal.

These rivalries are upsetting alliance management habits and leaving Europeans feeling "like bystanders, overtaken by events directed by others," in the words of Jens Suedekum, a leading European research scientist. French president Emmanuel Macron puts the same thought bluntly. Europe, he says, "needs a Sputnik moment"—a wake-up call that brings new resolve and dedication to excellence in science and technology.

These are some of the notable ideas aired in our expert essays and roundtable conversation on Europe.* Like the other regions we surveyed, Europe has had to contend with a technological revolution that has spread around the globe in ubiquitous and instantaneous fashion, leaving governments and other institutions scrambling to keep pace. In some areas, the European experience has provided examples to be followed. In others, failure or inaction has been the dominant European response.

Russia's decision to use cyber platforms to undermine democratic governments in Europe (as well as to interfere with the 2016 US election) also reflects a major transformation of global power politics. "The important conflict today is centered within states, rather than pitting states directly against each other," added one roundtable participant. In what seems to be a global trend, and one that crosses forms of government, what modern-day America Firsters "say about nationalism is similar to what guys in Hungary, Italy, or Russia say about that subject."

---

* We are grateful for our colleague Jim Hoagland's moderation of, and observations on, those discussions on Europe, the text of which forms this chapter.

## Digital Challenges from Russia, China, and the United States

Social media, by giving voice to the extremes of political discourse, has played an outsize role in fracturing national unity and undermining traditional political parties and parliamentary governance in the world's democracies. Traditional consolidated political parties in Europe have also been hollowed out, no longer the dominant sources of fundraising and event organization in politics. Unable to deliver these spoils, the parties have lost both influence and an ability to command a consistent platform or narrative among their members, as seen in the debilitating debate over Britain's referendum-mandated exit from the European Union. The United States saw a version of this in its own 2016 presidential primaries and election.

In such cases, it was noted, Russian president Vladimir Putin was not handicapped by a respect for the truth. This helped his generals to exploit a "gray zone" (below the threshold of armed conflict) of information warfare and develop doctrines to guide the buildup of a "hybrid" arsenal in Europe, mixing information operations and cyber capabilities alongside new kinetic armaments. Meanwhile, Washington and its allies in NATO have been slow to share information or develop joint responses to possible cyberattacks on national or alliance power grids and other forms of infrastructure, in the collective judgment of the participants in our project discussions. "We have not yet developed even an implicit form of the mutually assured destruction doctrine to protect ourselves," said one participant.

Russia has called attention to the creation of a new branch of its military that is dedicated to information warfare, the Brookings Institution's William Drozdiak noted in his paper covering the effect of social media, advanced technologies, and artificial intelligence in Europe. Moscow appears to have unleashed these units to stir trouble among Russian populations in Latvia and Estonia, as well as interfere with British, US, and German elections. President Macron's suggestions that NATO and Russia take the lead in working out an international cyber code of conduct have not been taken up by these or other nations. Countering the current cyber threat to elections is as much a governance problem as it is a technological one, with discussants noting the general successes in avoiding disruption enjoyed by an organized France in 2017 and by the United States in 2018.

Russia has not in fact developed significant digital platforms of its own. It has instead adapted Western platforms to its (largely subversive) purposes, at a relatively low cost. The cost factor is important for the economically weakened regime of President Putin. It is likely Russia will now concentrate more than it has on artificial intelligence and combine it with social media platforms to create deepfakes, using falsified photos as well as text. Putin, it was said by one expert on Russia, sees himself as the leader of a growing "illiberal internationalism" that reaches out to nativist political forces inside European nations. He seems to have convinced himself that Western intelligence agencies pioneered the use of information warfare by social media in Ukraine in 2013, and that he is simply paying them back.

These reinforcing technological, demographic, and political changes arrive at a time when NATO, the most successful multinational alliance in history, is experiencing intense internal strains as well. The Trump administration's America First policies have intensified Europe's long-held fears of US isolationism, while Trump's Washington sees major European powers as free-loading parasites that must be forced to change their ways. Until now, Europe has been thought of as a force multiplier for American strategic objectives in the protection of common values. But US bases on Russia's perimeter do not in themselves protect allied interests against cyber operations. A major reappraisal of the nature of the American-European partnership would be occurring now even without the America First political turn.

Relations between the United States and Europe are also perturbed by China's insistent push to make its technology dominant in European markets and to acquire key infrastructure assets such as a port outside Athens (acquired as payment for loans), infrastructure projects in Portugal (sold to China for $12 billion needed to pay off European creditors), and industrial technology firms based in Germany (though the sale of one toolmaker was finally blocked by Berlin). American fears that Chinese technology will be at the heart of Europe's development of 5G telephone networks has led to threats from Washington to stop sharing intelligence with Britain and other European nations if they were to use the Huawei corporation's cheaper 5G network equipment.

The third outside power preoccupying Europe's leaders is none other than the United States. America's tech giants have, in European eyes, smothered the continent's chances of developing its own national champions. And the Trump administration's disruption of negotiations for a Transatlantic

Trade and Investment Partnership, a favored project of German chancellor Angela Merkel, has left the allied nations without a clear framework for future dealings on advanced technologies.

## Economic Causes and Effects

Rather than encourage their companies to compete directly with Facebook, Google, Amazon, and others, European governments have chosen to regulate and tax American tech firms. Regulation and rule making are part of the EU's DNA, and these attributes have hampered its Single Digital Market strategy to build a unified tech marketplace for its 500 million inhabitants. Despite years of discussion among bureaucrats in Brussels, there continues to be no appealing way for consumers in Italy to buy French goods online, and language is a hurdle for information exchange. Only a unified digital market could provide the kind of scale and financial clout that have helped American and Chinese tech giants grow ever bigger. Experts at our roundtable observed that of the world's top twenty digital companies, none are European; of the top two hundred, only eight are.

The division of responsibilities and powers between national governments and the supranational structure of the European Union creates a unique blend of strengths and weaknesses in dealing with such startling technological change. So does the force of Europe's diverse and deeply entrenched cultural patterns of behavior and attitudes, which makes disruption a far less welcome force than might be the case in Silicon Valley. Also, the European Union is in large part a political creation formed through the adoption of rules, standards, and other bureaucratic imperatives—rule making, more than a constitution or shared sense of identity, is what binds the continent together.

Europeans have also demonstrated a greater attachment to tradition and to privacy than have Americans when it comes to new technology. When Google sent cars around Germany with cameras to map streets without advance notice, German citizens vociferously objected. Digitization of literature created an uproar in France.

Moreover, established businesses and sources of finance have a shared vested interest in making it difficult for newcomers to compete and prosper, as Caroline Atkinson pointed out in her research paper. An initial failure is not forgiven or overlooked in Europe as easily as it is in other societies. Many European would-be entrepreneurs migrate to Silicon Valley; twenty-three American billion-dollar "unicorn" technology startups as of 2018 had a European-born founder or cofounder, about one-quarter of the US total, and just shy of the number of unicorn startups within Europe itself.[12] Venture capital is a less important source of investment in Europe than elsewhere, too; flows have increased in recent years but are only about one-sixth the US volume.

Europe has seen more traction in its efforts to apply antitrust standards and to enact corporate codes of privacy, such as its General Data Protection Regulation (GDPR). The GDPR has been accepted by many of the large American firms, but as discussants pointed out, doing so may—counterintuitively to European entrepreneurial interests—simply be in the best interest of only large tech incumbents, given the cost and technical complexity of compliance for smaller firms. EU members have also found it easier to limit the role of political advertising in traditional and social media than is the case in America. So while European efforts at regulating tech have had unexpected dynamics, some in the United States now argue that there are useful lessons for US privacy and antitrust policies. The challenge for Europeans is to find ways to foster innovation.

Another bright spot is that the generous European safety net facilitates acceptance of disruptive technologies into the workplace. European employers have limited the fallout from the introduction of automation into manufacturing, which occupies a larger role in the overall economy than do services, representing a reverse of the American pattern. Rather than lay off excess workers, European companies tend to keep them on the payroll while retraining them for different jobs—often in the services sector. This is made possible by general educational practices such as those enshrined in Germany's apprentice system, which balances occupation-specific training with general education courses that are partly subsidized by the government. The value of this general education foundation may be a surprising finding to many Americans, who are largely more focused on reproducing the (also successful) applied-training aspects of apprenticeship models. This pattern does, however, exert a downward pressure on wages, meaning that EU countries in the late 2010s had record-high employment rates and relatively stagnant wages.

By 2018, German industry had installed eight robots per thousand workers, compared to a Europe-wide average of four per thousand. The comparative US figure was two. Jens Suedekum reported that while each US robot installed replaced three or more American workers, the net effect in

Germany was actually flat: while the manufacturing sector lost two hundred seventy-five thousand jobs alongside rapid robotization from the mid-1990s to the mid-2010s, this loss was more than offset by an equal job increase in the German business-related services industry. It is unclear if that equilibrium can be maintained as artificial intelligence begins to play a larger role in German manufacturing.

Discussion also explored the role of additive manufacturing—popularized under the label of 3-D printing—in reshaping Europe's role in the global-supply-chain system of manufacturing, which has become the key instrument of international trade during the era of globalization. Like American ones, EU firms heavily offshored industrial production in the 1990s. Now customers demand specialized products that are not well suited to distant mass production. So far, shifts in this direction are small in the context of the overall supply chain, but they are moving quickly. A fully automated manufacturing facility that Adidas established in Bavaria is being closely watched in Germany as a possible harbinger of a move toward reshoring.

## Demographic Realities

Europe is in any event confronting a future in which labor shortages will likely be a more pressing problem than labor surpluses. European countries are setting records for low fertility rates. If current trends hold, each new generation will be replaced by one two-thirds its size. (Gains in longevity will offset some of these losses, but not significantly.) Without a major increase in immigration, Europe's population could decline by one hundred million in the second half of this century, essayist Christopher Caldwell observed in his report to the conference. But resistance to immigration is growing more intense across Europe as social media fans the flames of cultural and religious animosities and populist parties vow to keep foreigners out.

A prosperous Europe does not control the causes of episodic waves of immigration aimed at its shores. Wars, deep poverty, and climate change are obvious triggers for mass movements of population. The drying up of Lake Chad, a combination of rainfall shifts and human mismanagement (which we return to in our next chapter, on Africa), was cited as one such trigger.

## Governance Lessons from the Emerging New World: Japan

In his essay on automation in Germany, panelist Jens Suedekum lamented the social and governance attitudes toward technology in Europe versus those in another economically and demographically mature US ally: Japan. How has Japan—which faces perhaps the most dire of the emerging new world's demographic realities, with a drastically shrinking workforce in conjunction with a rapidly aging population—reacted with such a constructive approach toward using that same new world's technological gifts to meet those challenges?

With a fertility rate that fell below the replacement level back in 1970, Japan's workforce has been contracting since the late 1990s, and the Japanese government expects that it will shrink by another 22 percent—or thirteen million workers—by 2040. And Japan's economy faces many challenges that would be familiar to Europeans: negative short-term interest rates, massive debt both government and private, and 1 or 2 percent annual economic growth.

Curiously, while Europe struggles with the core question of immigration as a partial response to such problems, Japan has largely dismissed it. Less than 2 percent of the Japanese population today is foreign born. Instead, Japan has focused on economic and social continuity through higher productivity.

By midcentury, two-fifths of its people will be over the age of sixty-five. But studies suggest that a 10 percent increase in the fraction of a country's older workforce is associated with a 3 percent decrease in total productivity.* Japan has therefore been focused on finding work that elderly people can still do, even when receiving medical services. Industries are springing up to provide elderly health classes, for example, in response to the government's actuarial realization that total social-health-care costs for aging are determined by how active or frail the elderly are. Annual physicals are mandatory. Meanwhile, Japan's broader "Abenomics" economic strategy demands and has been having success in bringing more women of all ages into the labor force.

Technological options are also emerging, with prefectural government-level subsidies for the use of robots not just in manufacturing but in services too. Sensors and robots in nursing homes are being used for communication and companionship with socially isolated residents, but they can also watch if people get out of bed at night and alert a skeleton staff.

Japan's closely entwined government and business culture, meanwhile, nervously watching China's own growing capabilities, has motivated a new strategy: "Japan, Inc.'s" traditional large export firms now work with smaller innovative firms, at home or abroad, to adopt their technology and integrate it into existing product lines, distribution channels, and customer bases. Stanford technologist Kenji Kushida described to us how they do so by quietly leveraging international offices, such as those established in Silicon Valley long ago to sell Japanese products to Americans, to partner with local start-ups. Large Japanese insurer Tokyo Marine, for example, is working with US start-up Metromile to spread its novel pay-as-you-go, usage-based insurance technology around the world. Or take Komatsu Construction, which has collaborated with a San Francisco–based drone start-up to take three-dimensional maps of construction sites and now offers that service alongside its core hardware products around the world. Similar examples play out across automotive, precision-equipment, energy, and consumer-electronic sectors. Meanwhile, Japanese firms are using their off-balance-sheet cash pile to enter limited partnerships with US venture capitalists and to sponsor potentially breakthrough university research, at home and abroad. Overall, Kushida recounted, the attitude toward new technology in Japan is very positive, since demographic realities mean that there is no concern about job loss from it. "No one is against robots."

One could say that the Chinese AI innovation model is government driven, while the Japanese AI model is solution driven. If the US model is, perhaps, tech driven, where does that leave Europe? A lack of workers in Japan is actually creating technological change, not the other way around, and that social transformation is being deeply internalized. To borrow a thought from Stanford socio-economist Karen Eggleston, the combination of changing demographics and AI is beginning to force a rethink in how everyday Japanese people structure their own home and professional lives. Japan's attitudes toward change show Europeans that there is more to navigating emerging new world than government policy.

* Nicole Maestas, Kathleen J. Mullen, and David Powell, "The Effect of Population Aging on Economic Growth, the Labor Force and Productivity," National Bureau of Economic Research, NBER Working Paper 22452 (July 2016), https://www.nber.org/papers/w22452; though, reflecting the paths available here, see also a competing argument that an aging society actually increases its productivity by responding to that threat with increased automation investment: Daron Acemoglu and Pascual Restrepo, "Secular Stagnation? The Effect of Aging on Economic Growth in the Age of Automation," National Bureau of Economic Research, NBER Working Paper 23077 (January 2017), http://www.nber.org/papers/w23077.

And so are demographic trends. Discussants observed that demographics are another realm in which Europe finds itself an early witness, but with little sense of control. While Europe's birthrates have declined since the 1960s, Africa's total population has grown 500 percent, to 1.3 billion. The seminomadic nation of Niger, which in 1950 had a smaller population than Brooklyn, will by 2050 have more people inside its borders than France will. In short, Caldwell warns, Europe sees a ticking population bomb across the Mediterranean, one that will be largely defused at Europe's doorstep. The median age in the EU is forty-three. In Africa, it is nineteen. The assimilation of young, unaccompanied immigrant males is considered a growing problem in the squares of some European towns; low European birthrates mean that indigenous children are already few, leading to fears of intergenerational cultural displacement. Small European towns have also been slow to attract so-called high-value migrants who possess needed skills or significant wealth.

Conference participants were not united on what steps Europe is likely to take to manage the social disruption that could accompany the growing immigration flows these numbers suggest (Figure 13). Countries like Sweden have welcomed so many immigrants as their native-born citizenry's fertility rates have declined that nearly 20 percent of the population is now foreign born, Caldwell said.

By contrast, cultural assimilation of immigrants into Europe's established societies has been relatively slow in other countries. "Frontier" countries such as Italy and Greece voice resentment over the numbers of refugees from Syria, Eritrea, and other war-torn societies they have had to absorb in comparison with more geographically removed European states. But, it was noted, even in Central European countries with relatively few migrants, public opinion polls show fear of and hostility toward the swelling ranks of refugees. The growing demand to move to Europe—one Gallup survey cited shows seven hundred million adults around the world reporting they want to move somewhere else, with 23 percent naming Europe as their preferred destination—is in sharp conflict with accelerating resistance to accepting immigrants. Our roundtable discussants also reached no general agreement on whether there is clear evidence of a strong link between immigration and crime in Europe.

There was, however, a general sense that the growing expression of social diversity produced by communications technologies and global immigration and demographic patterns presents the world with a governance challenge that can be satisfactorily managed by the democratic institutions of America and Europe if attention is paid and creative thinking applied. Caldwell argued that Europe has not treated the problems with either attention or creativity. Instead, the default position has been to expect the newcomers to adopt the European host population's values and habits automatically, or to be satisfied to live in sordid isolation in ghettos on the edges of big cities. One participant argued that European governments have done relatively little to educate potential refugees and economic migrants before they leave Africa or other points of departure about what awaits them and how their behavior will be expected to change in public settings in particular; the potential for sexual harassment, for example, has been a rallying point for emerging anti-immigration European party platforms.

Uneducated and untrained immigrants will find it difficult to take advantage of Europe's institutional assets, such as the German apprentice-worker system. Europe therefore has significant incentives to direct development aid and other forms of educational assistance to potential immigrant groups, it was suggested, with goals of either better preparing those who do leave for life in their new countries of residence or equipping them to stay where they are and prosper.

But that must be done with care. When President Macron of France touched on greater educational opportunities for women being a crucial step toward reducing fertility rates, he was criticized in Africa and elsewhere for, at a minimum, having a patronizing attitude toward Africans. Macron's predecessor Charles de Gaulle once observed that geography is history. The sense from the discussion was that demography is Europe's future, for better or for worse.

For more:

- "Europe and Technology" by Caroline Atkinson
- "European Demographics and Migration" by Christopher Caldwell
- "Europe's Challenges in an Age of Social Media, Advanced Technologies, and Artificial Intelligence" by William Drozdiak
- "Europe in the Global Race for Technological Leadership" by Jens Suedekum

*https://www.hoover.org/publications/governance-emerging-new-world/winter-series-issue-219*

## From "Europe in the Global Race for Technological Leadership"
*by Jens Suedekum*

Labor supply is decreasing. This can create skill mismatch, i.e., a coexistence between job displacements on one end of the market and labor shortages on the other end. But a prolonged "technological mass unemployment" is unlikely.

Japan, where the population is aging even more rapidly than in Europe, seems to have realized this. The country is deliberately forging ahead in introducing robot technologies, even in industries such as elder care. Hysteria about robots taking away jobs is mostly unheard of in Tokyo, but not so in Berlin, London, or Paris.

Why is Europe so concerned? Probably the key reason is that Europe doesn't see itself as the center of technological development anymore. It fears being overtaken by others and thus is mostly concerned about negative consequences of technological change. . . .

The five most valuable companies in the world by market capitalization are Apple, Amazon, Alphabet (Google), Facebook, and Microsoft. In the top ten, there are eight American and two Chinese firms. Europe has none. In modern software and information technology, Europe is almost entirely dependent on the United States. The same is true for military and defense. . . .

On a global scale, there has been a productivity slowdown across mature economies, with an average rate of increase of only 1.1 percent in the period 2008–15, compared to 2.2 percent in 2000–2007. Productivity growth strongly varies within Europe. But in all major European economies, it has been consistently lower than in the United States over time.

It may have been such observations and statistics that led Tim Höttges, CEO of German telecommunications provider Deutsche Telekom, to conclude that "Europe has lost the first half of the game called digitization."

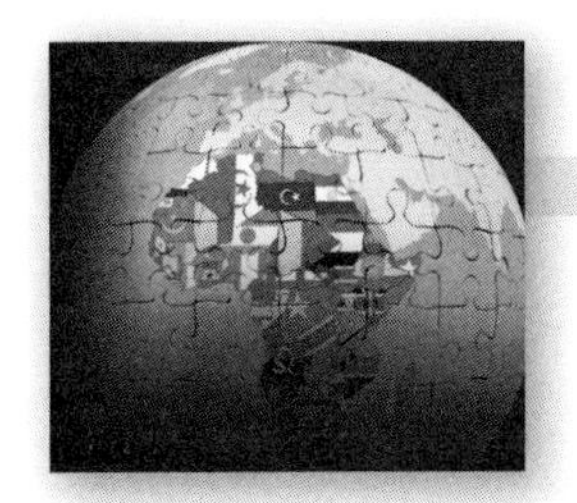

CHAPTER 6

# Africa in an Emerging World

The papers and discussions on Africa brought into stark relief Africa's current reality and the dramatic transformations that are already being driven by demography, climate change, and technological innovations. Our roundtable, guided by the experience of former US assistant secretary of state for African affairs George Moose,* also highlighted the enormous challenges these transformations will present. Together they paint a picture of Dickensian contradictions, with both bad news and good news, and reasons for both hope and deep concern.

The most disturbing news is the possibility for the forces of the emerging new world, which will be powerful disrupters globally, to have even greater disruptive impacts across the African continent. Population and climate issues arguably pose the most concrete threats to economic sustainability and social and political stability. Both could be managed through better governance—rule of law, responsive political representation, less corruption, and investment in human capital—but it is a risky strategy to expect the coming decades to offer simultaneous revolutions in African governance, when the previous ones were unable to deliver.

Roundtable participants flagged two other forces that are also at play in this continent and that will compound and amplify the impacts. First, the spread of extremism, religious and ideological, and its violent expression, which is driven and exploited by both local and external actors. And second, the rise of populism, nationalism, and authoritarianism, which further complicates the challenge of developing rational and concerted responses.

* We are particularly indebted to Ambassador Moose, vice chair at the United States Institute for Peace, for his authorship of summary observations from those sessions on Africa, from which the text of this chapter is largely drawn.

**FIGURE 14 Nigeria is on track to surpass the US population by 2050.** Nigeria is by far the most populous country in Africa, representing 15 percent of today's total. Due to persistently high fertility rates, most of Africa's largest countries are projected to at least double their populations by midcentury. Africa's rapid growth will put new demands on today's weak educational and social infrastructure. How Africa manages those challenges has economic and political implications for the rest of the world.

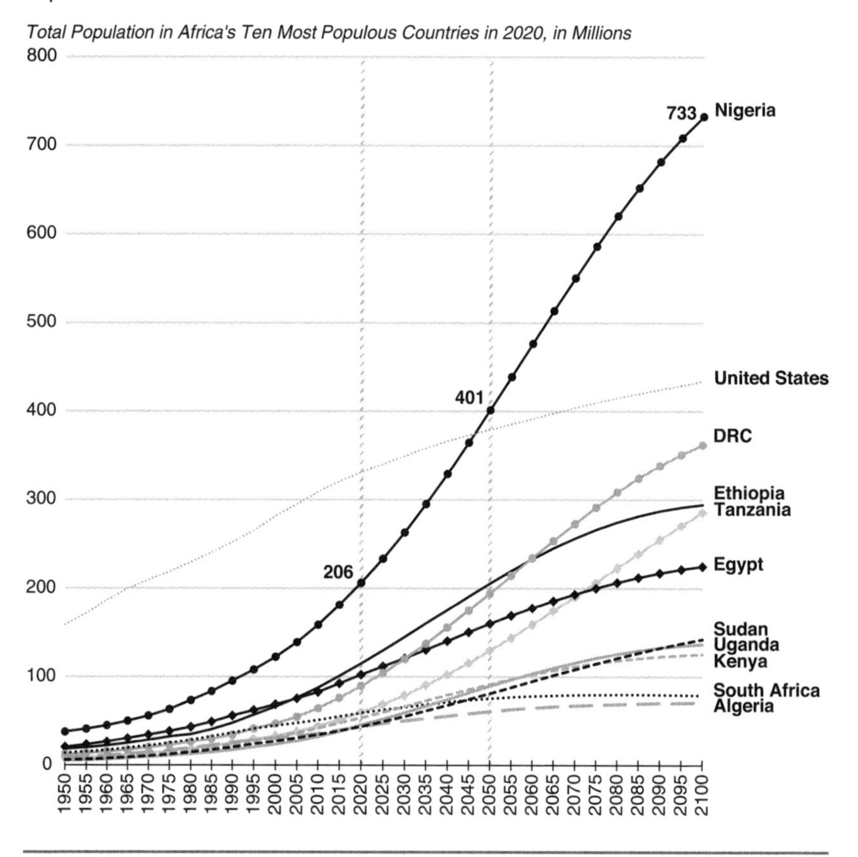


Data Source: United Nations, *World Population Prospects: The 2019 Revision*, medium variant. Demographic analysis by Adele Hayutin, 2020.

Taken together, these disruptions create huge governance challenges at all levels—national and subnational, regional, and global—implicating both governmental and nongovernmental institutions. Africa in recent years has seen many successes, with rapid economic growth rates in some states prompting labels of an "African century." And emerging technology could continue to fuel that, as societies are able to avoid some of the up-front costs of having to replicate infrastructures and norms that were crucial to modernization elsewhere in the world. At the same time, the scale of these changes also threatens to overwhelm the fragile capacities of governments and other institutions to manage their impacts.

The good news is that destabilizing outcomes are not foreordained. As with the influences that preceded them—industrialization, for example—these global changes have the potential to produce enormous opportunities and advances. Discussants also drew attention to policies and strategies that could mitigate the most destructive effects of these trends, increase resilience and adaptive capacities of governments and societies, and maximize the prospects for improved economic, social, and political outcomes.

Former US assistant secretary of state for African affairs Chet Crocker argues that the critical factor in determining whether these coming changes will yield positive or negative outcomes is the character and quality of governance—the ability of political institutions to make wise decisions and sound choices. In Africa, too, avoiding disaster and reaping the potential benefits amid this emerging new world will require thoughtful, focused, and concerted actions. Thoughtful and focused because there is a need to ensure that energies, resources, and investments achieve maximum impact. Concerted because the magnitude of the challenges far and away exceeds the capacities of individual actors. And they will require responses that go beyond the whole-of-government approach of coordination across agencies—especially given the weakness of many of these public entities in Africa today—to embrace more of a whole-of-community approach, requiring both governmental and nongovernmental focus. We can describe the challenge as one of governing over diversity. But it could also be described as governing in adversity.

## Climate and Demographics

Demographic and climate-based developments pose the greatest threats to the economic sustainability of Africa's fifty-four countries, as well as to their political and social stability. The dramatic impacts of these forces are compounded and amplified by the ways in which they interact with one another. Moreover, the basic trend lines driven by these forces are largely set. The people who will cause the next surge in population growth have already been born; and the climatic forces that will drive environmental change over the next fifty years have largely been put in motion by the patterns of the past fifty years.

As sociologist Jack Goldstone underscored to us, Africa's demographics are exceptional among global trends. Whereas improvements in mortality, education, and economic development have caused fertility rates in parts of Asia and Latin America to drop to about replacement levels (2.1–2.2 births per woman), Africa contains the world's only large swaths with fertility rates above 5.0.

The projected consequences are stunning (see Figure 14). Whereas world population growth outside Africa is projected to end within this century, Africa's population is projected to grow from 1.2 billion in 2015 to 2.5 billion by 2050 and 4.5 billion by 2100: from 17 percent of the world's population today to 26 percent by 2050 and 39 percent by 2100. The dramatic nature of this change is illustrated by Nigeria, whose population is projected to grow from 181 million in 2015 to 411 million in 2050 (more than the United States) and 794 million by 2100 (more than Europe).

This projected population growth is driven by high fertility, which, as we have said above, is itself a product of economic, cultural, and educational factors. Sub-Saharan Africa has a long tradition, for example, of strong extended family structures and high fertility. But the crucial factor here for our purposes is the lack of secondary education for girls. When young women are educated, they can exert more control over their own family size. It is the critical factor to reducing fertility.

Population and climate impact one another. Environmental scientists Mark Giordano and Elisabeth Bassini stressed that, while there is much uncertainty and great diversity, the outlook is for temperatures to rise faster in Africa than they will globally. As a result, the continent will likely experience longer, hotter heat waves, more protracted periods without rain, and more intense precipitation when it does fall, with negative consequences for arable land and agricultural yield. One consequence is that climate-related internal migration pressure will likely increase the movement of people from farms to cities due to diminished agricultural opportunities. Migration out of the continent will likely have a small impact on Africa's population, but the implications for receiving countries (namely in Europe) could be substantial, and politically explosive.

There will also be significant health impacts due to changes in the extent and location of disease vectors such as mosquitoes and the increased threat of diseases that spread from animals to humans. These impacts will in turn have consequences for both life expectancy and the quality of the labor force, not to mention health-care costs.

More than 600 million people in sub-Saharan Africa lack electricity, a number that has actually increased in absolute terms in recent decades as the rate of population growth outstrips the rate of new electrical connections. The average sub-Saharan African household consumes less electricity each year than a typical US refrigerator does, and over a million people on the continent die from indoor air pollution resulting from noncommercial household energy sources. Meanwhile, satisfying the energy needs of this growing population will cause carbon dioxide emissions and air pollution to rise—which can be moderated if Africa turns to its plentiful wind, solar, hydro, and uranium resources. The continent's natural gas will play a crucial transitional role, especially where it can displace coal and oil or indoor biomass burning.

## Climate and Conflict

In their paper on climate, Giordano and Bassini demonstrated scholarly prudence and caution about asserting a causal connection between climate change and conflict. However, even without drawing a direct connection—and from the perspectives of roundtable participants who had visited or inhabited the continent over the past forty years—we can already see examples of a nexus between climate and conflict. It is especially evident in the Sahel, where the steady decline in arable land was a major contributor to the conflict that erupted in 1989 between farmers and herders in the Senegal River basin, which escalated into a larger conflict between Senegal and Mauritania. It also exacerbated the interclan conflict that led to the civil war and ensuing famine in Somalia in the early 1990s.

For another example of the interplay already underway in Africa between the climate and institutional management of resources, Lake Chad has shrunk over the past half century by over 90 percent, while the population living around the lake has nearly doubled, from 17 million to almost 30 million. The resulting scarcity and unequal distribution of natural resources have fueled instability and violent conflict throughout the Lake Chad Basin, which includes northern Nigeria. Meanwhile, BBC World has documented the increasingly volatile swings between droughts and floods in northern Mali that are inflicting lasting damage on crops and livestock, exacerbating tensions between farmers and herders from different ethnic groups.

## Demography and Economics

Africa's population growth threatens to outpace the capacity of governments to manage and mitigate the social and political impacts; under projections for some countries, population growth actually outpaces economic growth. Importantly, a growing gap between these two measures could result in the poor being unable to improve their lot while economic gains accrue to the small economic and political elites. As has been observed elsewhere through this project, automation of manufacturing can accelerate the returns to invested capital over labor, which could cause the small economic and political elite of Africa to get even richer. As this happens, it will be important to find other ways for these technologies to make the poor richer, too.

Africa will have an excess of young workers, who could become a great resource within the continent, and who will also be in short supply in the rest of the world. After 2040, working-age populations will shrink everywhere in the world except sub-Saharan Africa. But if education does not improve (high school completion rates are 31 percent for men and 24 percent for women in the region), most African youth will not be able to compete with workers in South Asia or North Africa.

Despite this dramatic population growth, Africa will remain the world's least densely populated continent. The crisis that looms is one that arises from the disconnect between the combined impacts of population growth and climate change on the one hand and the lagging productive capacity to educate, feed, house, and care for the continent's rapidly increasing population on the other.

## Technology and Economics

With a growing middle class, Africa is becoming an attractive investment destination, particularly for the United States and China. Private investment and remittances sent home from migrant workers and the broader African diaspora both exceed development assistance.

As investor and technology evangelist André Pienaar described in his observations, Africans are embracing mobile technology as an enabler for greater education, better access to services, and improvements in people's lives. African mobile internet penetration (35 percent) is below the world average (54 percent) but growing rapidly. The technology sector accounts for an increasing portion of the growth in African economic output, with strong trends, for example, in Nigeria and Kenya. These activities have facilitated the growth of informal markets of citizens serving the needs of citizens, and mobile money-transfer systems have simplified payments for goods and services, creating new livelihoods. Supported by the growth of telecommunications and internet access across Africa, services make up 53 percent of GDP, with services exports growing six times faster than merchandise exports.

Opportunities exist to use technology to improve agriculture and manufacturing. Agriculture employs 60 percent of African workers and produces roughly one-fifth of its GDP. Improving productivity of food production and processing will be necessary to support growing populations while coping with the effects of climate change, disease, and drought. Manufacturing represents another fifth of African GDP. As Tony Carroll noted to us, drawing from his long experience in African business, the focused application of AI has the potential to improve productivity and overall production in both sectors.

Internet connectivity provides young entrepreneurs with access to information, partnerships, and capital, but many countries struggle with poor communications and internet infrastructure. Prospects for the technology sector are also adversely affected by education shortfalls, especially at the secondary school level, and by the fact that educated professionals leave Africa for higher-paying jobs and better quality of life elsewhere.

Chances are further dimmed by serious institutional and governance deficits. Capital markets are weak. Government reform is needed to establish predictable and transparent investment environments, which are also critical to the development of the technology sector. Historically, we have seen, in

Africa and elsewhere, concerted resistance from established institutions—e.g., state telecommunications monopolies—to policy and legal reforms that would undermine their positions, market control, and revenues. That resistance to change is almost always aided and abetted by governments and political elites that also benefit from existing monopolistic regimes and that may also feel particularly threatened by these new technologies and their potential political uses. Discussants pointed to notable exceptions where the technology sector has shown impressive growth. Kenya, despite its other serious political and governance issues, appears to be one example. It would be helpful to have a better understanding of why that is so. Meanwhile, in Sierra Leone the literal collapse of the state created space for the expansion of cell phone technology.

Sub-Saharan Africa remains the most underconnected region of the world. What little trade there is with the rest of the world remains in raw commodities, mostly agricultural and mineral products. Intra-Africa trade remains small; it favors manufactured and consumer goods; and it is limited by weak physical and human infrastructure, small individual country markets, tariffs and other barriers, absence of trade finance, currency risk, corruption and rent-seeking, and civil disruption. Technology alone cannot solve these problems.

## Technology and Governance

Access to the internet by Africa's youthful and urbanizing population provides information and power to individuals and groups, which can be used to make governments and markets more transparent and accountable and institutions more responsive to citizens' needs. But it has also made it easier for networked activists (and criminal, trafficking, and terrorist networks) to overturn fragile African states. This in turn can increase the determination of leaders to seek tools of political control.

Mobile internet technology is being used by public-private partnerships of Western governments and local NGOs to support peace, for example by amplifying messages of unity and cohesion over division and discord, enabling peaceful and transparent elections, and countering the use of social media for recruiting by terrorist groups. But it is also being used by indigenous actors to sow discontent, manipulate public opinion, stoke racial and ethnic tensions, and discredit and delegitimize political leaders and governing institutions.

Increasingly, technology and social media are also being employed by external actors. China is making major investments in technology, intentionally aligning state interventions with those of the Chinese private sector, including giants Huawei and Alibaba. With those investments comes influence, and questions about how that influence will be utilized and to what ends. Chinese companies affiliated with the CCP are building and managing the transition from analog to digital broadcasting in much of Africa, as they are in parts of Latin America. Meanwhile, there is evidence that Russia is using the same tactics it has deployed in Europe and elsewhere to manipulate elections and public opinion. André Pienaar is not alone in his concern that Africa is becoming the arena for a global contest between digital democracies and digital autocracies.

## Governance and Leaders

To quote Ambassador Crocker, "In Africa, as in every region, it is the quality and characteristics of governance that shape the level of peace and stability and the prospects for economic development. There is no more critical variable than governance, for it is governance that determines whether there are durable links between the state and the society it purports to govern."

In the relatively new African nations, the challenge for leadership is to build a social contract sufficiently inclusive to allow effective governance over diversity. In addressing this challenge, Africa starts out with a significant deficit in the form of weak social compacts between states and societies, which is not surprising given the continent's colonial and Cold War history. Governments that have come to rely on foreign counterparts and outside investment in natural resources, rather than domestic taxation, have weaker connections to their citizens.

Building more-inclusive governance systems and political processes becomes infinitely more difficult when the landscape of who needs to be included is changing or expanding rapidly and dramatically. The youth bulge has revolutionary potential, with implications for political and social instability and the possibility of conflict. The different impacts of these shifts on various regions, subregions, and groups only add to the difficulty of managing them.

These changes and the other challenges presented in these papers place extraordinary burdens on Africa's leaders. It is worth recognizing that human agency—the leadership of men and women in official and nonofficial roles—will be at least as decisive as the abstract variables and vectors described in these studies. Wise leadership is central to the building of inclusive governance. Africa's external partners bear a parallel responsibility to support wise leaders and to nurture the institutional legacies they help create.

African governments and institutions will not be able to overcome these challenges and deficits on their own. They will need outside help. External actors, however, seem increasingly incapable of (or opposed to) exercising stewardship and mobilizing global capacities in this area. Traditional sources of leadership—the United States and Europe—are preoccupied with their own internal governance crises and challenges. At the same time, global trends favoring populism, nationalism, and authoritarianism are undermining respect for and promotion of democratic values and good governance. In consequence, global institutions (namely, the United Nations) are further weakened and marginalized in their ability to act and exert influence.

Meanwhile, the rise and growing engagement of undemocratic international actors (e.g., China, Russia, Turkey, Iran, and Saudi Arabia) are amplifying governance challenges. Behaviors of some major external forces, such as China, may contribute to short-term improvements, but over the longer term they will constrain growth possibilities and exacerbate economic and social dislocations.

The global context is further shaped by a larger security dynamic: the concern over global terrorism. Cold War considerations led states to prioritize security issues over issues of good governance, and to overlook—or worse—the flaws of those African political leaders deemed to be important in the confrontation between East and West. An eerily similar dynamic is at work with respect to the global war on terrorism, with Western governments inclined to downplay the extent to which support for the activities of African security forces, and the governments that nominally control them, aggravates governance issues. There is also the question of whether Western assistance, however well intended, serves to support incumbent regimes rather than reform flawed governance systems. In our public aid and in our private investment arrangements with Africa's political elite, we would do well to remember that the West has the global power not just for intimidation, but more importantly for inspiration.

The global pushback against liberal governance norms has consequences in Africa, as emboldened authoritarian governments act to close the space for civil society to operate. Illiberal centers of power and great power polarization reduce the influence of Western and other traditionally democratic societies and provide opportunities for African states to move toward authoritarian, state capitalist policies. Western aid and investment, while not perfect, has traditionally come alongside liberal frameworks meant to encourage local governance transparency, pluralism, and the rule of law. That carrot is less enticing alongside less-disciplined offers from China or Russia.

It is too soon to tell whether states can evolve toward inclusive agendas or face fundamental tests of strength between social and political groups. Success could depend on developing the economic and financial resources required for including various social groups and demographic cohorts in the development of governance institutions.

If a critical mass of large and influential states (such as South Africa, Nigeria, Kenya, Ethiopia, Ivory Coast, and the major North African states) head in a positive direction, they will pull some others along. If more leaders were to practice inclusive politics, the outcomes could be better. The reverse is also true: if the abolition of term limits, patrimonialism, and kleptocracy become regional norms, it will be harder for better-governed states to resist the authoritarian wave.

One positive force that must not be ignored or underestimated, and that may ultimately determine the future, is the growing popular demand for better governance and accountability, a trend affirmed by the continent-wide polling of AfroBarometer. The question is how best to enable and empower civil society to institute and measure governmental accountability.

Finally, as one insightful roundtable participant pointed out, it is unrealistic to expect a revolution in governance in the near term. For the foreseeable future, good governance will remain the rare exception rather than the norm. Therefore, it is important to consider what can be achieved in an imperfect governance environment. Outside of Africa, Bangladesh offers one example of a country with similarly deep population, environmental, and resource challenges, plus a historically weak state—but where a benign government has nonetheless managed to avoid hampering an enthusiastic domestic and international civil society and a hardworking population. Kenya may be another.

## Governance Lessons from the Emerging New World: Bangladesh

In the words of social scientist and director of the Subir and Malini Chowdhury Center for Bangladesh Studies at UC Berkeley Dr. Sanchita Saxena, "What makes Bangladesh successful?" Here is country of 160 million that since the 1960s has been a symbol of a seemingly inescapable quagmire: dire poverty, tens of millions of people living below sea level and subject to annual monsoon flooding, few natural resources, dangerous urban working conditions; and yet—atop this new hinge of history, still poor and facing a host of new challenges, from climate change to dislocation and global trade disruption—"successful." As Bangladesh takes its place among middle-income countries, what can its experiences in playing from a poor hand teach African countries with similar prospects?

According to Saxena, the country has seen successes through its continuous, if gradual, improvements in public health, education, and income generation. International nongovernmental and multilateral aid groups have been a critical piece of this progress, but they have intervened in other struggling countries around the world too, with less positive results. Rather, progress came down to the political will of successive Bangladeshi leaders to recognize peoples' needs—as well as their own institutional and resource limitations—and craft a strategy that could be called a sort of benign negligence.

Citing the work of Dr. Naomi Hossain,* she argues that this was the consensus among the country's elite political and business class, which, much like the Western liberal international commons born of the Second World War, was forged through trauma, in this case a bloody war for independence in 1971 and an ensuing famine in 1974 that killed millions. The social contract that emerged was to provide minimal protections to major shocks, especially to rural women long seen domestically as the enablers of development (witness the global phenomenon of women-driven rural microcredit, born here, or women's overwhelming role in the ready-made garment industry). Pro-market reforms were undertaken in order to receive international aid and investment. And, in turn, both main political parties hoped that the electorate would then grant legitimacy to their governments as reward.

In this hands-off framework, domestic civil society has taken the lead. Bangladeshi domestic group BRAC is actually the world's largest NGO, among hundreds of other social development, religious, and even media groups delivering the sorts of social services one might expect of a state.

One example: grassroots, door-to-door rural messaging campaigns to educate mothers on the most effective diarrhea treatments for children, over time reducing deaths. Bangladesh has also dramatically reduced fertility, through a combination of education and contraceptive programs, from over six births per woman at the time of independence to just replacement rate today. Both efforts were funded by the international community, but with the interventions themselves delivered from the ground up, relying on locals to spread the message. One area where the Bangladeshi government has constructively focused its energies through this process is in the collection of unbiased data and statistics on the country. That social report card is transparent and acts as a sort of public infrastructure to encourage both NGO activity and foreign direct (or local) investment. It's a potential model for African nations with similarly willing leaders but serious public resource constraints.

What of the future? The Bangladeshi garment industry dominates the country's private sector, representing three-quarters of total export. It underpins today's economic growth, but it is threatened by both inherent flaws in global supply chains and the emergence of new automation technologies such as 3-D printing and sewing robots, which could make it easier for global buyers to shift supply chains to higher-cost-of-labor parts of world. It is not given though that new technology will simply displace Bangladesh's role on the supply chain. Some argue that Bangladeshi labor is so cheap, given its relative productivity, that it will be hard for machines to replace. And the Bangladeshi government has developed new infrastructure such as roads and ports to help make the country a low-cost option in logistics as well as in labor.

Notably though, Bangladeshis could respond to the changing world by securing their own place in it: local producers would be particularly well served by investing up the value chain, acquiring new technologies themselves and then leveraging their existing trade relationships to profit from that. In a country where labor violations persist, investment in productivity could also improve working conditions. It may well be the world's first buyers of automated "sewbots" will not be resurgent textile mills in North Carolina, but rather the manufacturers of today in the developing world who already have the requisite industry relationships and customer bases. If Bangladesh is successful, the real concern may be for the next in line—namely, those in Africa with an exploding workforce over the coming decades—who could see their own future place on that cost-of-labor ladder boxed out by a nimbly adjusting incumbent.

* Naomi Hossain, *The Aid Lab: Understanding Bangladesh's Unexpected Success* (Oxford, UK: Oxford University Press, 2017).

One cannot consider these points on Africa's future and not come away with a powerful sense of the need for urgent action. There are equally compelling reasons for citizens and policy makers in the United States and elsewhere to act: humanitarian concern for the fate of present and future African generations, but also prudent regard for their own national interests. But, facing such an array of challenges, the United States must focus our limited energies and resources and prioritize those actions and interventions that seem to be most likely to achieve the maximum impact. With that in mind, the following list is intended to be suggestive rather than definitive.

## Education

If there is one across-the-board theme that rises to the top of the list of priorities, it is the importance of education, with a pronounced emphasis on secondary education, especially for girls. The single most important driver of high fertility rates, especially in sub-Saharan Africa, is minimal secondary education for girls. Expanding access to that education, combined with the employment possibilities that flow from it, is essential to lowering fertility rates and beginning to flatten the curve of Africa's explosive population growth. It is equally essential for preparing Africa's rapidly growing youth population for productive work and employment, which in turn is a critical underpinning for a stable and civically engaged middle class. The development and expansion of a more educated populace is in and of itself a driver of economic growth, creating both a demand for and an ability to supply new products and services.

Alongside education, increased mobile connectivity and broadband access can help empower women and lower fertility rates. Educated women, primarily urbanites, ultimately will have to lead public-education campaigns to reduce fertility and modify family structures. Their efforts will be made much easier if young women with fewer educational opportunities can gain access to mobile technology and, with it, greater knowledge and exposure.

In this context, health and health education are critical adjuncts to education, both in changing attitudes and behaviors that affect fertility and in supporting healthier populations. Retaining and attracting the continent's most talented and highly educated youth may prove to be even more difficult and daunting than educating them in the first place.

## Agriculture

There was also a strong consensus on the need to improve the productivity and sustainability of the agricultural sector, given the proportion of the population involved (60 percent), as well as the need to increase production to feed the continent's rapidly growing population. Results of these efforts will depend on closely related interventions, including investments in both physical infrastructure (e.g., roads, transportation, and storage facilities) and institutional infrastructure (e.g., extension services to educate farmers on current best practices, marketing support).

Issues of land tenure and land reform, which are both legal and cultural, pose a daunting challenge to this effort. No less daunting is the question of how to promote the expansion of industrial agriculture while at the same time protecting and preserving the position and contributions of traditional small-hold farmers. Models do exist for developing complementary relationships between traditional and modern agriculture, but success will depend upon policies based on an understanding of those complementarities.

## Technology and Economy

Technology, including AI applications, could be the critical enabler for expanding educational opportunities and assisting in the adaptation of the agricultural sector. But its potential contribution is much greater, extending to both the industrial sector and commerce in general. The question for both government policy makers and private-sector entrepreneurs is how to harness technology in ways that maximize Africa's growth potential. The answer to that question depends on success in tackling some significant existing structural and legal problems, including how to break down existing governance structures (e.g., state monopolies) that impede growth and the effective and efficient use of resources, and how to address the problem of weak and inconsistent laws and law enforcement (including reform to data collection and data privacy policies).

Going forward, Africa will also need to create a policy and investment environment that incentivizes technological innovations, specifically those in automation and artificial intelligence that could rapidly improve productivity, and the wireless and internet infrastructure to facilitate business and commerce of all sorts. Harnessing the continent's demographic dividend will also mean developing a workforce that can create and apply those tools through the uptake of science, technology, engineering, and math (STEM) curricula in primary and secondary schools, as well as to online platforms.

Two other challenges stand out. The first is reducing barriers to intra-African trade that severely constrain possibilities for economic growth. The second is finding ways to channel diaspora remittances and external investments, which together far exceed Official Development Assistance (as defined by the OECD), to productive economic activity. In some areas, the informal economy is the primary source of employment and opportunity. Yet today, external investments, such as the high-profile Chinese initiatives, are directed largely into the formal economy, such as real estate projects, which yield few secondary benefits for broad-based economic growth. Remittances from the diaspora, on the other hand, flow primarily to families and into the informal economy. They are one of the largest sources of foreign-resource inflows into Africa and a source of great opportunity for the continent.

## Climate

Important steps can and should be taken to prepare for climate change, including more research about a continent for which there is paltry data on climate trends; more investment in agricultural research to develop such tools as adapted seeds and farming techniques (think Israel and its successful fight against water scarcity) and improved soil and water management; the strengthening of health systems; and assisting African governments in developing plans and strategies to mitigate and adapt to changes in climate.

As just one example of such adaptation, discussants speculated on how the dramatic cost and time reductions in the genetic engineering of plants enabled by emerging CRISPR/CaS9 technologies might help spread the gains in yield and survivability seen in cash crops elsewhere to the indigenous orphaned crops of Africa that were passed by in the global agricultural productivity Green Revolution of the 1960s. Importantly, the investments that can be made now in increasing Africa's resilience to the climate challenges it already faces will become only more valuable on a warming planet.

Meanwhile, there is an urgent need to radically expand and scale the development of clean energy of all sorts—and not limited to the small-scale-distributed renewables on village microgrids (e.g., solar light bulbs). For example, given the growing global commercial attention toward the continent's indigenous natural gas resources, investors and local governments should deliberately channel efforts into cultivating practical domestic uses for that gas, to reduce energy poverty and to support local economic development, rather than viewing it as an opportunity for export earnings. In homes, natural gas–derived propane burners could displace dirty indoor biomass stoves and potentially avert hundreds of thousands of indoor air pollution deaths annually. On a larger scale, sorely needed fertilizer plants could become "anchor tenants" on indigenous natural gas pipelines and distribution grids, upon which other businesses could be further developed. This is a clear opportunity for US government involvement given its own growing influence and geopolitical interests in the global liquified natural gas trade.

Not least, in the environmental arena, there is also a need for strategies to protect and preserve Africa's forests and carbon sinks. These efforts have both local and global benefits.

## Governance

The logical place to begin is with reinforcement for good actors, those governments attempting to move in the right direction (Ghana, Ethiopia, Senegal, Benin) and those political leaders who have demonstrated a commitment to change (Ethiopia, Angola).

Also important is strengthening support for responsible and legitimate civil society actors and using both leverage and incentives to enlarge the space for democratic engagement. That includes arming civil society actors with the technological tools to promote transparency and accountability, both political and economic, as well as enabling the use of mobile internet technology and social media platforms to counter drivers of conflict and promote peace (see the work done, for example, by the Washington, DC–based accelerator PeaceTech, which brings promising African entrepreneurs to the United States to meet investors and be introduced to scalable cloud-based data and computing platforms that can fuel their growth). Better protection of property rights complements these efforts, encouraging entrepreneurship, decreasing hostility to outside influences, and promoting inclusiveness and trust within communities.

Implicit in the roundtable assessments and discussions was a recognition of the need for much greater focus on boosting the capacities of urban governments and institutions, since it is in Africa's rapidly growing urban centers that the greatest governance challenges will arise: meeting the needs for basic services (education, housing, employment, health) and creating structures that allow for the management of the inevitable competition, tension, and conflict that arise in urban settings.

Better regional and intra-African institutions will also be crucial, especially in addressing climate change. Weak regional governance and interstate cooperation have hindered past efforts on water and other resource management. The Okavango Delta is one example; the Mono River project that was designed to serve Ghana, Togo, and Benin in West Africa is another. At the global level, it has become increasingly clear that the United States and others will need to develop ways to counter Chinese and Russian exploitation of internet technologies and social media platforms and the havoc they can cause in the realm of governance. At the same time, the United States and other external forces should examine carefully the consequences, intended or inadvertent, of their own polices and interventions. Discussants observed that the United States has distinct global advantages in and experience with a variety of twenty-first-century technologies—for example, cloud computing, fintech, and cybersecurity, as well as digital entrepreneurship and governance norms—and it should consider how to best use those to meet a growing appetite for such services across Africa.

The issue of Africa's growing debt burden is brought into sharp focus by China's often exploitative lending practices, which have saddled African governments with unsustainable debt burdens and in turn made them hostage to China's political wishes. There is an irony in the fact that China is actually one of the greatest beneficiaries of responsible international lending as established by the World Bank and other global lending institutions. This underscores the fact that the debt issue is a matter of global governance, one in which international lending institutions, backed by concerned governments, need to intervene to constrain exploitative lending practices.

Efforts to improve African governance must be undertaken, but with full recognition that success will be halting and will differ across states. Our contributors therefore probed the challenge of finding ways to improve economic and other outcomes even in the absence of good governance. Experience in other parts of the world suggests that, where national leadership proves unresponsive, important gains can be made by working to buttress governance at the state, municipal, and local levels. In addition, more can be done with mobile technologies to strengthen civil society organizations and their ability to hold governmental authorities accountable, and to expand the space for both political and economic participation. The same technologies and social media platforms can create opportunities to channel remittances and other financial resources into productive, job-creating investments, which can further make room for nongovernmental action.

As grim as the projections and prospects may sound, this is not an excuse for resignation and inaction. There is nothing inevitable about the most dire outcomes.

However, if good outcomes are to prevail over bad ones, affirmative action and effective management—i.e., good governance, or at least the lack of bad governance—will be required. Thoughtful, focused, and concerted interventions can make a difference.

The encouraging news is that we have examples from history of the international community's ability to respond to major challenges and threats. There is, of course, the Marshall Plan, a model in its own time for responding to crises. But a more recent and relevant example may be the international community's response to the HIV/AIDS pandemic, which had Africa at its center.

The awareness of the crisis came in the late 1980s, when epidemiologists were able to document the evolution of the epidemic and its potentially catastrophic implications, not only for health but also for economic growth and social and political stability. Throughout the 1990s, the US government, among others, struggled to mobilize an appropriate response, hampered by its own political and cultural controversies taking place within its own borders. It was not until 2001 that a response emerged, the PEPFAR (President's Emergency Plan for AIDS Relief), authored by the George W. Bush administration. It was followed shortly by the creation of the Global Fund to Fight AIDS, Tuberculosis and Malaria, which also benefited from strong political and financial support from the Bush administration. Some twenty years later, sparked by these two global initiatives, the international community has mobilized more than $80 billion to combat the spread of AIDS, much of that directed to Africa. In consequence, HIV/AIDS transmissions have been dramatically reduced and health outcomes greatly improved.

Across the hinge of history, Africa may have yet another model ahead—the will and capacity of its citizens or kin abroad to reengage with the continent. One roundtable discussant, a Ghanaian expatriate now working in the Silicon Valley tech industry, described how twenty-first-century communications technologies and network platforms dramatically reduce the costs to members of the African diaspora to connect and coordinate with other family members and local champions back home. This powerful global network will only grow as African emigration does. And as it does, it offers a way for those on the outside who have long been concerned with the future of this continent from abroad to more effectively harness the incentives and structures—political, economic, or social—that are in play there, in direct cooperation with those who understand them best. This has potential to change the ground game on which the emerging new world's Dickensian forces of population, climate, and technology will play out in Africa.

The existence of deep and dynamic challenges is not new for Africa, which has shown great resilience throughout its jagged history. The most recent example derives from the early months of 2020: whereas developed Western countries largely found themselves flat-footed in their initial response to the COVID-19 pandemic, many health authorities in Africa, where infectious and parasitic diseases already claim a third of all lives, were quick to recognize the gravity of the threat and to take early isolation measures despite (and perhaps in recognition of) very poor resources ultimately available to treat it. This shows the capability of the continent under duress.

So there is still time to answer these questions. And in an emerging new world, there are new tools, and therefore new hope. But with urgent demographic and climate challenges ahead, the clock is ticking.

For more:

- "Africa Trade and Technology" by Anthony Carroll and Eric Obscherning
- "African Governance: Challenges and Their Implications" by Chester Crocker
- "Climate Change and Africa's Future" by Mark Giordano and Elisabeth Bassini
- "Africa 2050: Demographic Truth and Consequences" by Jack A. Goldstone
- "Unlocking the Potential of Mobile Tech in Africa: Tracking the Trends and Guiding Effective Strategy on Maximizing the Benefit of Mobile Tech" by André Pienaar and Zach Beecher

*https://www.hoover.org/publications/governance-emerging-new-world/winter-series-issue-119*

**From "Unlocking the Potential of Mobile Tech in Africa: Tracking the Trends and Guiding Effective Strategy on Maximizing the Benefit of Mobile Tech"**
*by André Pienaar and Zach Beecher*

The failure of Western governments, especially the United States, to get involved in funding virtuous applications of social media and mobile technology to democratize dialogue and enable entrepreneurship in Africa risks allowing the vital African market to fall into the hands of strategic adversaries. The innovation economy in the United States and the United Kingdom is often compared to that of China, but it is important to grasp that the Chinese model is a radically different proposition than the Western model. The application of cybertech and AI to internal controls means that China and its close allies are building technology-based authoritarianism. This is in sharp contrast to our open systems and the underlying altruism of our innovation economy that enables our freedom of choice, even if the luxury of having so many choices and so much convenience can at times be overwhelming to us.

However, the impact of social media and mobile technology also requires meaningful commitment to the virtues at the heart of Western liberal democracy. Central to this is the role of good governance, which is the essential building block to a strong and lasting peace and to the promise of economic growth. The potential success of technology must offer a tantalizing incentive to compel leaders to offer the good governance required to allow and enable the continuing flourishment of a sector in which millions of Africans are placing their hopes for the future. For venture-capital investors in Europe, the Middle East, and Africa, corruption and innovation are opposing forces. Where one is present, the other is invariably absent. Innovation thrives on good governance. Corruption corrodes and kills innovation and opportunity, thereby wasting the talent of generations. Standing at the edge of a transformative era in technology and the economy of Africa so as to better determine its own destiny, African governments and their Western partners must demand transparency and a strong commitment to the values of the liberal international order to enable their citizens to flourish in the digital domain.

PART

# An Emerging America

So far, we have considered the impact of global transformations on major countries and regions around the world and on democratic processes. We have explored how artificial intelligence, advanced manufacturing (particularly 3-D printing), and the other technologies of the "Fourth Industrial Revolution" will transform the world economy. Global supply chains may shift as we learn to produce goods closer to where they are consumed, and AI-enabled technologies may change the relationship between workers and machines, thereby redefining the workplace. At the same time, we are witnessing dramatic workforce demographic shifts. Although we cannot forecast the exact character of these transformations, history teaches us that they will likely be bumpy. It also teaches us that they will create new opportunities for prosperity and human flourishing. And, fortunately, the United States is well positioned to take advantage of that potential.

Now, we will look closer to home, to the future of the US economy, society, and government in a rapidly changing world. We asked experts from around the country to consider what advancing technologies and demographic transitions will mean for our **domestic economy and workforce**, and they proposed ways for the United States to manage those impacts and take advantage of new opportunities to ensure a growing, productive population. Beyond appreciating the gravity of the problem, what can be done at different levels of government and enterprise, given the social license for it?

Improving productivity underlies economic growth, and it is here that emerging technologies present their most compelling story for the United States. We start with an assessment of the future labor market by MIT's Erik Brynjolfsson, using a novel data set from the professional social network LinkedIn. In his research, Brynjolfsson identified skills from today's jobs that may be performed in the near future by machine learning, finding that machines can do at least some tasks of almost every job, suggesting we will see substantial redesign of work and significant reskilling across the economy. Some occupations, such as retail, are likely to be disproportionately affected. He also considered the degree to which an individual employee's skills and education have value to his or her employer—the more valuable, the larger the incentive for the employer to directly invest in its own employees' upskilling. To us, Brynjolfsson's findings show the value of enabling bottom-up, interest-based frameworks in approaching what is ultimately an individually tailored challenge. This is an area where the United States can excel—and if we play our cards right, it additionally offers the prospect for big-time improvements in productivity.

The United States grew up with the premise that individualism and ingenuity create opportunity and prosperity. But recent decades have seen backsliding. First is economic calcification, such as the spread of job licensing or housing restrictions, that has made it less rewarding for people to switch professions or move to more suitable parts of the country. Then there is the compromise to the idea of the individual itself, subsumed instead into a group-based identity founded on common characteristics—age, race, gender, class, profession, political beliefs, or so forth—each with a set of inherent blinders and limiting norms. Making the most of the great technological opportunity now emerging will require bringing back flexibility in America, defined broadly. Drawing from the scholarship of a variety of economists, including John Cogan and Michael Boskin, this will mean both flexible institutions and flexible people.

It's rare in the policy world for experts to agree on something. It's even more rare for them to do so consistently over time. But through the course of our study, one thing has emerged as the absolute core of enabling personal flexibility. And that is education. In America, our K–12 education system is a failure in plain sight and has seen little improvement despite massive new federal investment over the past five decades. Hoover fellows Macke Raymond and Eric Hanushek organized a project session here specifically on this topic, drawing upon a lifetime of research to explain how America needs to focus more on improving the results of our K–12 education—the children—and less on the conditions of the adults that are part of the system. The lesson from Germany, probably the world leader in technical and workforce education, is that the key to being able to retrain someone later in life—individual flexibility, in other words—is rooted in their having received a solid basic education in the first place. That solid initial education teaches people how to learn, something that sticks with them long after they have forgotten, for example, the date of the Norman Conquest. Improving American schools will be a requirement of crossing this hinge of history right side up.

And while the United States dominates the global ranks in elite colleges and universities, not every American can attend—or wants to attend—four-year institutions after high school. A changing economy and workforce will require an adaptive educational system, and Van Ton-Quinlivan, former executive vice chancellor of the California Community Colleges, explained how America's generally competent, widely distributed, locally directed, and low-cost community colleges can contribute to preparing workers for twenty-first-century jobs. Drawing from her experience at the helm of California's overhaul of its community college–based workforce-training programs, she outlined how other relatively nimble higher-education institutions across the country can support continuing training and education for a modern economy.

What about when things don't go as planned? Flexibility also matters when it comes to government or even private-sector support programs. We should think differently about our social safety nets if we believe that the changes brought by the emerging new world might see more Americans drawing on them at various points in their lives. We embrace the technological changes of the emerging new world, but ensuring program effectiveness in a faster-moving economy is part of our country's duties toward those who may get hurt as society overall advances, and we share our project contributors' thoughts to this end.

Then there is the **security** situation, and America's role in it. The security commons that has had the United States at its core since the Second World War has not been absolute, but it has put an umbrella over the domestic affairs of many parts of the world for long periods of time. The technological changes that are now occurring in each of these countries and regions, however, directly affect the nature of the US defense posture as well.

Namely, the United States finds itself increasingly engaged in strategic competition with both China and Russia, even as its technological edge erodes, and, though great-power competition may dominate the conversation, the US military continues to operate in Afghanistan, the Middle East, and Africa. At the same time, we live in an age of rapid innovation, with artificial intelligence, additive manufacturing, advanced computing, and other technologies enabling new military capabilities and changing how wars will be fought. So we studied how these emerging technologies affect the strategic and operational dynamics in the two theaters of great-power competition—the Pacific and the Eurasian landmass—and with regard to nonstate actors.

Former supreme allied commander Europe General Philip Breedlove (USAF, ret.) and Georgia Institute of Technology professor Margaret Kosal reviewed Russia's traditional approach to innovation and considered how major emerging technologies, including advanced manufacturing and materials, might be employed by Russia, the United States, and the latter's NATO allies. Looking to the Pacific, meanwhile, Admiral Gary Roughead (USN, ret.), along with Emelia Spencer Probasco and Ralph Semmel from the Johns Hopkins University Applied Physics Laboratory, saw that information dominance will be central to US-China competition. New technologies, from AI and autonomous systems to space capabilities, are changing the nature of conflict, requiring a rapid evolution of both military technology and operational concepts. Finally, National Defense University's Colonel T. X. Hammes (USMC, ret.) argues that emerging technologies give nonstate actors military capabilities traditionally only available to major powers, thereby shifting the balance of power in their favor. The United States must reconsider its old assumptions and carefully redefine its strategy for dealing with nonstate adversaries armed with these new capabilities.

As retired USMC general and former secretary of defense James Mattis has described the prospect of conflict where computer algorithms are given the lead, "Think of a war that is not limited by fear."

CHAPTER 7

# Emerging Technology and America's Economy

Classical approaches can work. That was the surprising message delivered by discussants at our roundtable on the interaction of emerging technologies with the US domestic economy. Education, migration, and responsive regulatory policy were all offered as examples of modest governance strategies that have worked before to help the United States economy take advantage of rapid changes while mitigating their disruptions.

It's tempting to frame rapid technological change as an unprecedented challenge for this country, and one requiring unprecedented forms of governance. But America has made assumptions like that before. Similar arguments were offered, for example, in favor of "new economic methods" such as President Nixon's draconian—but roundly praised—economy-wide wage and price freezes during the attempt to deal with the unexpected inflation of the early 1970s. What sounded to many like a bold and sorely needed change ended up failing spectacularly and sent the US economy on a decade-long downward spiral. Our discussants therefore warned against throwing out orthodox policies for untried alternatives, as the result of doing so would be to replace one set of uncertainties—the complexity of the coming change itself—with two.

## Productive Work

Erik Brynjolfsson, the widely published economist and technologist from MIT, describes machine learning as potentially the most important technology of this generation. Three things drive its rapid ascent: 1) the mass digitization of data throughout the economy, more closely linking the data sets of our computers with the environment of people and our daily activities; 2) significantly better computing power that reduces the time needed to run computationally expensive machine-learning decisions by orders of magnitude (and more closely matching human decision time frames); and 3) better algorithms that give better or faster results given any set of data. These parallel changes allow machines to effectively share work with humans, with each assuming the tasks it does best.

In his contributions to this project, Brynjolfsson quoted Stanford AI pioneer Andrew Ng in expecting that "anything today which can be done by a human in less than one second is well suited to be done better by a machine." Since 2015, for example, machines have surpassed (a general population of) humans in image recognition. Since 2017, voice-recognizing machines have approximately equaled human abilities on call switchboards. Applications are proliferating. When eBay in 2014 introduced fully automated, machine-learning-based language translations to its listing titles in the Latin American marketplace, US exports to the region's buyers increased by 11 percent, essentially overnight. So machine learning can improve the functioning of existing markets.

While some applications have burst onto the scene very quickly, raising governance questions in the process, Brynjolfsson explained why potentially more fundamental changes will actually take some time. The brake is not so much the technology itself as it is the need to redesign jobs, since some tasks in most occupations are suitable for machine learning while others will continue to require human labor.

To that end, Brynjolfsson's comments aimed to tease apart one of the key governance questions around emerging technologies: the impact on employment. He described machine learning as being different from earlier types of automation, and it is not possible simply to extrapolate from previous experience to understand this new field. He described how machine learning would cut across wage and skill levels, with lower-wage jobs disproportionately affected. But most jobs will be reinvented rather than eliminated.

First, Brynjolfsson argued the importance of seeing today's jobs as bundles of activities and tasks. A radiologist, for example, can be said to perform twenty-seven distinct tasks as part of her work, some of which can be done better by machine learning, some of which cannot. A minority of radiologists of the future may spend more time teaching machines to do those particular tasks well. Most radiologists will spend less time on those tasks compared to other important parts of their jobs, such as patient interaction. Redesigning jobs will be key to machine-learning productivity gains.

Secondly, employment is not a matter of slicing up a pie. The supply and demand of jobs and the tasks they comprise are dynamic. While machine learning and artificial intelligence can substitute for some human activities, they can also augment them. Returning to the radiologist example, consider that a doctor using machine learning to augment her diagnoses could potentially diagnose more patients in a day, in fact making her a more valuable (that is, productive) employee than she was previously. Supply-and-demand elasticities could play a role as well, with hospitals able to offer more radiological services at lower prices, and consumers then deciding to take more CT scans. Finally, there is the potential for new, utility-enhancing tasks that might emerge through invention and reengineering—using radiological machine learning to perform wholly new types of diagnosis and monitoring, for example, or automated remote care that simply wasn't available before, and so on.

Finally, Brynjolfsson argued that machine-learning job impacts can be predicted, or at least understood, by using a skills-based framework. This breaks down the problem into manageable pieces: we can enumerate the skills required in each job, and we can separately evaluate the actual progress of machine learning or other potentially disruptive technologies on each of those skills. This does not dictate a single-policy response, but it does create a useful map to guide good governance efforts for policy makers, who are understandably concerned about effects on their constituents and want to prioritize their efforts where they are most needed. And it should give the confidence to freely encourage the pursuit of the productivity-enhancing aspects of these technologies, rather than a defensive crouch in an attempt to prevent the impacts from arriving on their own. It's important to remember that future upside, which does not otherwise get to advocate for itself in today's policy making.

A country's wealth is directly linked to the growth of its workforce plus the growth in its worker productivity. Through this project we have seen how US demographic trends are relatively healthy compared to those of other world powers. But troublingly, growth in worker productivity has stagnated since the 2008 recession. Emerging technologies such as machine learning or additive manufacturing are promising antidotes to this, but they must first be applied in the existing economy. To that end, our discussants described how US business reorganizations, which are an indicator of firms' becoming more efficient to make use of new technologies or market conditions, are actually happening slower today than they were twenty to thirty years ago. And the incorporation of machine learning into American businesses may be no different, with an estimated ten-to-one ratio for in-firm worker reskilling and reorganization costs compared to actual investments in machine-learning IT. Up-front costs to such upgrades are high, with productivity gains following later. Such trends are not new: electricity, for example, was an obviously excellent technology, but it took thirty years for factories to see productivity gains from it, given the need to strand existing (often steam-powered) assets while developing new skills, methods, and business models to fully take advantage of the benefits of the new technology.

This points to the need to encourage not just entrepreneurship, which is always welcome, but also, more broadly, an efficient labor market.

How to do this? Updating worker skills will be important, and applied education is a topic to which we will return. But part of an efficient labor market is the matching of skills and capacities to the needs of employers. As architect and urban planner Alain Bertaud has observed, in prison everyone has a job, and a short commute at that, but you could not argue that convicts are living their most productive lives. Flexible markets create the structure for such matching to occur naturally—through an emergent order that makes use of individuals' knowledge about themselves, their capabilities, and their preferences, as well as the workforce needs of industry—in ways a government planner never could. Moreover, a number of government rules and regulations in place today already reduce employer-employee flexibility. Our project's discussants offered a few concrete recommendations to drive improvements in US labor markets given coming disruptions.

One was for a reversal of the explosion in occupational licensing, which has grown to cover 25 percent of all American workers, up from just 5 percent in the 1950s. Extensive licensing requirements for trades, many of them in the service sector—whether yoga instructors, physicians, or fruit pickers—began with the justification of protecting public health and safety. Amid the decline of US union membership, however, licensing is now used by trade lobbies of existing workers in a well-paying profession as a means to explicitly block qualified new workers (who they fear might over time put downward pressure on wages) from entering their field. Licensing makes it harder for someone to start a new line of work without unnecessary time and expense.

Until recently, for example, Maryland required fifteen hundred hours of training to work in a blow-dry hair salon. Arizona required one thousand hours. Since this licensing is generally state-based, it can also restrict worker mobility in some sectors, weakening the great diversity and expanse of the United States labor pool—one of our unique global attributes.

Improving that mobility across jobs, especially across geographies, will be another key challenge. Americans have long been a mobile people, and the idea of setting out for new fortunes in a new town or state is part of our psyche. And Brynjolfsson's work suggests that machine learning will have varying geographic effects across the country. But Americans have become less mobile. In 1970, the Velvet Underground reflected a long and deeply held American dictum when they sang, "I do believe, if you don't like things you leave, for some place you've never been before." Since that era, the share of Americans who had moved in the previous year to a different location within the same county has fallen by half, in a steady decline. The share that had moved from one county to another has declined by more than one-third. And the percentage of young adults—generally the most mobile segment of the population given the need to establish households and start careers—who had moved at all in the previous year has also fallen by one-third. Too many Americans are staying in areas or jobs that are not demanding their full potential.

Our discussants described some of the theories of why this reduction in mobility may be happening, and potential governance strategies to address it. High student-loan debt, for example, may be keeping even well-educated young people at home living with parents. Census results show that the share of young adults (ages twenty-five to thirty-four) living with their parents stayed steady from 1990 until the mid-2000s but exploded thereafter, from 11.6 percent in 2005 to 22 percent in 2017. This is a bid to reduce the cost of living, but it may also reduce these young adults' potential to gain income from finding the best jobs available to them, with career-long earning impacts.

Another factor may be the prohibitive cost of real estate in the nation's most productive urban regions, such as the San Francisco Bay Area; Los Angeles; Seattle; Boston; Washington, DC; and New York City, which now disproportionately produce jobs in excess of the suburban or rural areas, which once dominated. Many vibrant US urban areas (with the notable exception of Houston) suffer from decades of housing-development restrictions and high construction costs that have suppressed the supply of new housing to well below job growth. This drives up rents and drives down home ownership (down by one-sixth among young adults since the year 2000). Some have argued that it also drives down marital rates (down to 40 percent from 55 percent over the same period) and, eventually, fertility, with long-term demographic implications. One recent study by economists Chang-Tai Hsieh and Enrico Moretti estimated that local zoning restrictions since 1964 have reduced the size of the US economy by one-half.[13] Now, one-half is probably an extreme estimate, but it points to the foundational importance of a well-functioning labor market, and how lack of labor mobility and occupational adjustments can encumber its efficiency.

If the flashy threat of labor disruptions from artificial intelligence and other emerging technologies is what it takes to prod policy makers to publicly revisit otherwise esoteric and special-interest-dominated policy topics like zoning or occupational licensing, then all sectors of the US economy could come out much the better for it.

And of course all of this process of technology implementation will take human creativity. Amazon.com reinvented the bookstore (and other stores) not by substituting machines for humans in shelving and checkout, but by changing everything in the supply-chain process from the supplier to the end consumer, down to the location and operation of warehouses that could offer millions versus thousands of products. Going forward, the process of creative destruction will require rethinking entire businesses around taking advantage of machine-learning technology. Brynjolfsson offered a startling prediction to technology bears: even if machine learning's progress froze today—say, no more announcements from Google's research teams or from Stanford's computer science labs—the US economy would still see decades of innovation on business practices that would improve aggregate productivity. Put another way, the much-publicized spread of artificial intelligence is in fact firms applying existing machine-learning technology to an ever-widening expanse of industries and problems. This innovation will require lots of intangible investments. Consider that Americans across a number of firms have already spent $100 billion on the development of self-driving vehicle technology, even though no driver has yet been replaced. Machine learning is not a loaded gun. In fact, the bullet has already been shot.

What, then, do Americans owe to each other in this emerging world?

## Better Education

First and foremost are educational opportunities that give individuals the most options to succeed. Universal primary and secondary education was a key governance response to the technology changes of one hundred years ago that brought Americans from farm to factory and urban life. And it was done with the awareness that society's ability to benefit from new capital investments in the means of production would require similar gains in its human capital.

A good education provides the foundation upon which individuals build careers and lives. It generates the human capital that is the engine of economic growth. And it makes for a stronger, healthier citizenry and, in turn, country. This has been true for generations, to the point that it is almost self-evident, but it bears repeating, particularly as we consider how new technologies and social dynamics are reshaping our society and economy.

Education, in particular during the formative years of kindergarten through high school, will only grow more important—and more determinative—in the world emerging before us. Unfortunately, America's K–12 education is lacking. As shown by our Hoover Institution colleagues Eric Hanushek and Paul Peterson along with Stanford's Laura M. Talpey and the University of Munich's Ludger Woessmann, there are deep and troubling problems in precollege education, problems that have endured for a half century.[14] Students from the high end of the socioeconomic distribution still outperform their peers from worse-off families by as much as they did fifty years ago. Moreover, though fourteen-year-olds learn more now than they did then, those gains are wiped out by the end of high school; a seventeen-year-old student today, be she rich or poor, is on average no better educated than she would have been in 1970. As the authors put it, "There is no rising tide for students as they leave for college and careers." This is a tremendous public-policy failure and a disservice to generations of students. And it is a failure that will hurt the next generation of students even more as they enter a rapidly changing professional and political environment.

Fortunately, thanks to the diligent work of many academics, policy minds, teachers, and student advocates, we know how to make our K–12 education system better. In our investigations for this project, former secretary of education for the state of New Mexico Christopher Ruszkowski put it well: "We have an implementation crisis, not an ideas crisis." The United States can improve its schools and support its teachers through transparency, flexibility, and accountability, and luckily the same technologies that are raising the importance of education may help improve it too.

## Why K–12 Education Matters

Let us step back and consider why education matters so much and why it will be such a defining policy challenge for the emerging world. Not long after taking office, President Obama visited Wakefield High School, outside of Washington, DC, and told the assembled students: "No matter what you want to do with your life—I guarantee that you'll need an education to do it."[15] That is no less true now.

It used to be that young men and women entering the labor force could reasonably expect to work only a few jobs over the course of their careers. They would change jobs infrequently and rarely move across sectors. But that is shifting. People will have to move between jobs more or, even if they keep their job, their tasks and responsibilities will evolve as new technologies come online. So workers will need to be retrained and to learn new skills. As we think about how to do that, we can take a lesson from Germany: recall Jens Suedekum's report that Germany has done a good job of helping workers change tasks or occupations because its citizens receive high-quality primary and secondary educations. A good K–12 education gives someone the fundaments of knowledge and learning that make them more agile later in life and better able to learn new skills.

To return to President Obama's message, the students of today will increasingly do many things with their lives, work many jobs, and take on many challenges, so they need a good education more than ever. But to further echo the former president, a better education is not just good for the individual—it is also good for the country. As he told those Wakefield High freshmen, "What you're learning in school today will determine whether we as a nation can meet our greatest challenges in the future." This is true in economic terms as well as in social and political terms. Artificial intelligence, 3-D printing, automation, and the like present a host of new opportunities for high standards of living, productivity, and economic growth rates. New technologies promise to expand the scope of what a single worker can do, whether through more dynamic robotic arms manipulating what humans cannot or through smart systems freeing humans from the drudgery of data entry. Erik Brynjolfsson's research has found that machine learning can do at least some tasks of almost every job, and he foresees "substantial redesign of work and significant reskilling." If workers can nimbly transition between roles and take on new challenges and jobs, we will all be better prepared to seize the opportunities presented by global technology advancement. We will see real growth.

It is worth noting here that we are not focused just on science, technology, engineering, and math (STEM) but on all disciplines. Yes, math whizzes, nuclear physicists, and other STEM graduates will help drive innovation and keep the high-end technology ecosystem humming, but we ought to be careful not to focus too much on STEM education. Reading, writing, critical thinking, teamwork, and social skills—all skills honed in primary and secondary schools—will remain invaluable for all Americans. Arguably, these "soft skills" will be even more important to workers in the near future; humans are uniquely able to perform tasks that require social and communication skills.

K–12 education helps individuals flourish and keeps the economy chugging, but it is also about the long-term health of our country. As we have said before, the world and the United States need good governance. We need to get governments back in the business of solving real problems, and we need to restore order to our democratic processes. We are covering no new ground to say that a well-educated citizenry will help do all that. Many have worried about the impacts of political advertising in social media or, more sinisterly, cyber-enabled influence operations on the preference formation of the American people, and have called for regulation or government oversight as the solution. But a better education would help US voters to rely on their own wits in navigating the forest of information presented by TV, social media, and other communications technologies, thereby helping them make more-informed choices in the voting booth. Basic civics education would give voters a road map for evaluating the plans and performances of political leaders. A well-educated population makes for a richer country, yes, but also a healthier democracy and better governance—in other words, a stronger country.

## What to Do

Improving K–12 education is a defining challenge for the United States, but how can we go about better preparing students for the emerging world? To answer that question, we held a roundtable at the Hoover Institution and solicited the input of leading policy minds and practitioners: our Hoover colleagues Eric Hanushek and Macke Raymond, the director of Stanford University's Center for Research on Education Outcomes (CREDO); the aforementioned Christopher Ruszkowski, former secretary of education for New Mexico; and Adam Carter from Summit Public Schools, a charter-school network. They represent a diversity of views, from the ivory tower, state capitals, and the classroom, but in conversation with them we reached a broad consensus.

What matters most is how much students learn, not measures of class time, spending, or other indicators. The two most important factors for student learning are their parents and the quality of their teachers. However, America's education system today is not designed to maximize student learning; it is too often oriented toward what is best for the adults in the system—be they teachers, administrators, or politicians—not what is best for the students. Fortunately, there are many successful schools around the country. They teach us that good schools collect data on student performance and feed it back to teachers and parents in a positive way, emphasizing what works, to foster continuous improvement, while lower-performing schools are afraid of data or unwilling to have that transparency for fear it will be used against teachers. Finally, education policy should focus on taking what works and doing it at scale.

Take that list from the top. Eric Hanushek and his colleagues' work on the education achievement gap, referenced above, compares how much students learn now as compared to in decades past. Student achievement is the focal point, and so it should be for administrators and education officials as well. Research indicates that the most effective way to increase student learning is to improve teacher quality. Students learn more when they have better teachers, and they in turn are more successful. We can all recall the impact a few good teachers have on us. But the reverse applies as well; in fact, Hanushek argues, the negative cost of a bad teacher is greater in magnitude than the positive impact of a good one. For a class of thirty students over just a single year, an above-average teacher (in the seventy-fifth percentile of teacher quality) increases the present value of the class's lifetime earnings by $430,000, but a below-average teacher (in the twenty-fifth percentile) decreases it by $800,000. The drag produced by poor teachers can be felt across America's K–12 education system. If you removed the bottom 5 to 8 percent of teachers, the United States would be near the top of the global rankings in education quality.

There is no magic formula for preparing high-quality teachers. Research shows that being a good teacher does not depend on particular certifications or backgrounds. But we can observe teacher quality and measure it through student performance and supervisor evaluations. In other words, hiring and paying teachers according to certain certifications and degrees will not necessarily improve the quality of teaching. Instead, because the effectiveness of teachers is revealed through performance, a primary focus of education policy should be assessing and maximizing annual improvements in student performance.

Experience tells us that improving teacher performance is not always easy, or possible. Some argue that teaching is a skill people largely have or don't have. If a teacher struggles even after a few years of experience, odds are he or she will not get much better. However, Chris Ruszkowski's experience as secretary of education for New Mexico is instructive here: to improve teacher performance, they began with "level-setting honesty." His office worked with schools to establish tools to assess performance truthfully and to communicate that assessment back to teachers and administrators. With a baseline set, it is possible to reward good teaching, help teachers improve, and enforce some measure of accountability.

Transparency of that sort is a hallmark of good schools, and they use it not as a stick but as a carrot. Routine assessments of performance and positive, constructive feedback can be found in many of the high-performing charter schools in America, for example. Those schools tend to have what Macke Raymond called "a culture of continuous improvement." They are focused on student learning and employ the tools available to them to assess how they are doing. Moreover, as Raymond's research shows, more than five hundred such high-performing charter schools around the country outperform both their state average and the local district schools in measures of student learning. Their success should motivate us to think differently about schooling: they have proven how to succeed and how to do so at scale.

This is not to say that charter schools are the only solution. Charters have a wide distribution of performance, with many failing to keep pace with the schools down the street or even the state average. Those schools' struggles should be a reminder that the trade-off of accountability for flexibility—the operating mantra of charters—requires accountability! Charters should be held accountable for their performance or face the consequences—and district schools should too.

When he entered the governor's mansion in Tallahassee, Jeb Bush prioritized improving Florida's education system, and his administration's success in doing so is a testament to the value of accountability and positive feedback. With the state's schools lingering near the bottom of the nation in performance, Governor Bush implemented a system to grade all schools on an A–F scale and direct funding accordingly. Schools were rated on student performance. As Governor Bush explained to us in the course of this project, his administration set the expectation that all students would reach basic levels of learning. Around the country, class time is set as the constant; Florida determined that learning ought to be. In the first year of reforms, 30 percent of third graders were held back, and teachers lost some of their job-security guarantees. Parents were given more choice in where their students went to school, and eventually over half the students went to schools their parents chose—including private schools through scholarship vouchers, public charter schools, public specialized curricula schools such as magnet programs, and community college dual-enrollment programs. Governor Bush bore a political cost by doing so—just before an election, no less—but he stayed the course, and the results speak for themselves. In a state where the majority of the students are minorities, schools jumped from below average nationally to near the top. They ranked sixth-best in the nation in math for fourth through eighth grades by the time Bush left office. Other states have accomplished similar reforms. Under Governor Susana Martinez, Christopher Ruszkowski set up an A–F rubric for New Mexico schools as well. He and his colleagues measured teacher performance and devoted themselves to rewarding good teachers through public acknowledgment—and financial compensation. They too saw student learning increase and schools improve.

You can see the theme: these reformers introduced new measures of accountability into ossified institutions by establishing broad-based cultures of transparency about school and teacher performance, and they used student learning as the yardstick. They focused on maximizing how much students learned, rewarding those schools and teachers that contributed positively and helping to improve those that did not. And as needed, they enforced their expectations, holding students back who could not read and do math on grade level, and addressing underperforming teachers.

There is another important element to boosting the overall quality of teaching in America: flexibility in hiring, training, and developing teachers. A good teacher does not necessarily show up with a traditional teaching certification. Private schools have the freedom to hire essentially anyone they like. And charter schools can similarly hire teachers with unusual backgrounds, meaning they can attract candidates from a broader pool of talent. Some charter or innovation schools have developed their own teacher-training and development programs. Summit Schools, a Bay Area–based charter network, is premised on the idea that the school makes the teacher. Summit's Adam Carter spoke to us about its model of teacher development. It sees teaching as "a set of skills, mindsets, and habits" that schools have to establish early on and sustain. Summit teachers go through a unique teacher-training program when first hired. Through hands-on experience and time spent with veteran educators, those new instructors learn a set of skills, mindsets, and habits designed to help them succeed. Summit Schools, like the other schools discussed above, preach transparency, regularly collecting data on student performance and evaluating teachers. And, again like so many other successful schools, they embrace that transparency as a tool for positive growth, not as a cudgel for punishment.

An important part of Summit's success, and that of any school, is the ability to recruit top talent. This is a recurring problem, in part because of how we compensate teachers. Eric Hanushek speaks of the bad equilibrium in teacher pay: we don't pay teachers well, and we get what we pay for. Generous and poorly constructed retirement plans are to blame, at least in part. Nine in ten public school teachers have defined-benefit pension plans, which were established decades ago and now drain state and local education budgets. Both teachers' unions and school districts are often incentivized to negotiate "Cadillac" defined-benefit retirement promises, since the value is high, but the people in the negotiation room won't actually have to pay for their full costs. Salaries, on the other hand, are due tomorrow. But the promises don't go away—eventually, when states and districts do increase funding for education, much of the new money goes straight to paying down yesterday's underfunded pension funds, not today's teachers. In Chicago, for example, one-quarter of state funding for education goes toward teachers' retirement benefits.[16] These pension funds are broke. The unfunded liabilities of the

California State Teachers' Retirement System (CalSTRS) total $100 billion. That is almost $3 billion more than the total amount spent on K–12 education in California last year.[17]

Meanwhile, across many districts, we can see the growing number of administrators and resource personnel of various kinds in schools—more adults but not more teachers. Every dollar for them is one fewer available for the teachers who educate our children. Likewise, costly efforts to decrease class size focus on getting *more* adults in the system at the expense of getting the *best* adults in the system.

With little money left over for boosting teaching salaries, teachers are paid less on average than what their peers in other industries are, so we get teachers who tend to be below average in college performance. This is an unsustainable system. A transition is possible, but it would require talking about what is good for the students as part of a policy conversation that has been largely about what is good for the adults. There is, perhaps, a grand bargain to be found that would trade better pay for compensation and performance reforms. This bargain could promise higher-wage salaries in return for defined-contribution retirement plans—as adopted decades ago by essentially every private-sector employer when confronted with the actuarial impossibilities of defined-benefit systems—and greater accountability and transparency in schools, including the forfeiture of certain tenure and job-security protections. It is not enough to simply assume that paying teachers more, making them happier, or adding more teachers will benefit a child's education. We need to recruit and retain high-quality teachers while throwing less money after the rest, or states and municipal governments will continue to see their ledgers go further into the red with little to show for it.

America's schools will not all become alike, nor should they. And they will not all become world-class. But we can keep working at the problem through incremental reforms, re-creating what works while jettisoning failing policies. Over time, small improvements add up. We do each generation of children a disservice by accepting poor-quality education as the status quo. As one participant put it, "When we have poor K–12 education, we are condemning people to lives below their capabilities." Instead, we should set high standards for schools and encourage them to be open about whether they are meeting them or not. We are fortunate to have available to us new means of collecting, storing, and learning from data more cheaply and quickly than before. Put those tools to use. Collect data on teacher and school performance and share that information with teachers and with parents to set expectations and improve performance. In other words, hold people accountable but also reward quality results. Secretary Ruszkowski described one school in New Mexico that outperformed expectations year in and year out—a straight-A school. They did well each year, but eventually the school board and administrators started grumbling that they did not receive any particular recognition or attention for their excellence. It is an understandable impulse, particularly because state education officials tended to focus on poorly performing schools, but an avoidable problem. Ruszkowski and his colleague visited the school and used what bully pulpit they had to praise it in front of parents and the local residents who fund it—a model for the kind of positive reinforcement we have in mind.

If we are transparent about standards and performance, and use that transparency to enforce accountability, then we can afford to let schools be flexible. We know how to make schools better. Moreover, a society as diverse as America's should not have a one-size-fits-all education monopoly. Washington cannot and should not dictate the education experience of a fourth grader in Fairbanks or an eleventh grader in Barstow. Only states and local municipalities have the resources or access to support transparency and enforce accountability across the thirteen-thousand-plus school districts in America.

There is a political cost to implementing reforms such as those described above. Jeb Bush has spoken of the risk he took in pursuing aggressive education changes in Florida. Had they failed, even if just in the near term, he would have been out of office in the next election, and his career would have ended. But he accepted the risks and put his political capital on the line. The demonstrated improvement in student performance became a political asset. His commitment should inspire education reformers on both sides of the aisle.

We appeal to government and unions not to see reforms as a threat or to fear loss of power or influence. The real threat is that the system founders with them at the helm, and it winds up our country's future with it. Detroit automakers—management and workers—were for decades happy with the deals they had arranged among themselves, until exposure to a changing world revealed just how little they had actually invested in their products in the meantime.

Improving the US K–12 education system will make America healthier, wealthier, and wiser. It will make us better able to weather the disruptions of technological and demographic change and to seize the many opportunities afforded by those changes. And it will make us better able to face up to this hinge of history and lead the international movement toward peace and prosperity.

Improving K–12 education should be a nonpartisan priority. It is too important now and will become only more so in the emerging new world. We need leaders to push for real change and to keep working at the problem, whatever the immediate political headwinds may be.

## Continued and Technical Education

Given a good K–12 foundation, we can do more. The choices that American students make about acquiring skills through their educations, and that workers make about learning new skills while on the job or between jobs, underpin what bundles of "tasks" they will be able to productively perform in a changing economy. Sixty-five percent of current US job openings require some level of skills acquired through postsecondary education, and our discussions of state and local community colleges as increasingly important institutions for providing applied education to a diverse spectrum of Americans supported that goal. Former vice chancellor of the California Community Colleges Van Ton-Quinlivan reported in her analysis for this project that five years after completing a two-year "career technical education" program at a California community college, a worker makes an average salary of $66,000. Five years after completing a two-year general education associate's degree, the average salary is just $38,500. At the same time, employers report seeking a broad array of general skills in addition to occupation-specific or technical skills—critical thinking, problem solving, language and effective communication, teamwork.

Education in an emerging new world will not be a matter of funneling students into today's "recession-proof" jobs (which may see novel challenges from emerging technologies), or of focusing on STEM education. Rather, the goal should be to produce graduates who have specific skills that meet the needs employers are looking for today, and a broad enough framework for overall learning that they can successfully return to the education system again and again throughout their careers to quickly acquire new skills as the task bundles change.

Why our interest in two-year community colleges? First, they already exist and go relatively unnoticed in policy dialogues that jump between the deep dysfunction of the American K–12 system and this country's relatively high-performing—but very expensive—four-year university system. Of the state's two million unfilled jobs, half require a four-year college degree, but half need less than that. And for students who are driven, community colleges often find themselves completing the remedial teaching that high schools frequently fail to deliver. California's 115-school community college system, our discussants noted, is likely the largest higher-education system in the country.

Second is the fact that in federal, state, and local governance environments, where budgets are likely to be increasingly crowded out by compulsory spending items such as health-care entitlements and pensions, community colleges remain focused and cost-efficient. They are often located close to home with yearly price tags of $3,000 to $6,000, versus many multiples of that for a longer (and often necessarily residential) four-year option. This is good for government budgets, and it is good for students to better avoid loan-debt traps.

Third is these colleges' track record at educating a diverse range of students. In California's two-year institutions, for example, 60 percent of students are women, 37 percent are 25 or older, 24 percent have children or other dependents, 31 percent are from families in poverty, and 64 percent work—40 percent of those full time. Their customer base more closely matches the profile and needs of midcareer students of the future.

Today's students seek a return on their educational investment: good jobs that will help them support themselves, provide for their families, pay their loans, and save for retirement. Community colleges with career and technical educational offerings are attractive to students for enabling them to enter occupations that are known to be productive and locally in demand—nurses, emergency medical technicians, welders, utility line workers, plant operators, or maintenance and repair technicians, for example—and their graduates (in California, 80 percent) can typically stay within their own region to find work. Local community colleges are able to partner with nearby employers, individually or perhaps more effectively in regional coalitions, and in doing so keep pace with their evolving needs by mixing and matching task-specific modules with general skills such as English, math, and social reasoning.

Our discussants noted the increasing importance of two-year community colleges remaining nimble as emerging technologies drive acceleration across the economy and society itself. California's system provides examples of agile programming, such as minimizing the bureaucracy that can slow the rollout of new curricula or pooling resources across smaller colleges to build effective collaborations with regional employers so that these employers see it in their direct interest to interact with students through teaching, internships, and curriculum development.

We also identified the opportunity for community colleges—with their relatively short "business cycles"—to more directly engage with both the employers and the high schools that will be providing their next crop of students, so as to reduce friction in the handoff. Employers can increase graduation rates, for example, by front-loading tuition reimbursement instead of paying employees back after they have incurred the costs. Or community colleges can expose their curricula and major options to high schoolers and their teachers to telegraph future career options. The latter idea was described as being particularly important for minority students who, once in the community college system, tend to select familiar but generally less productive (and lower-earning) areas of study: in California, the top major for Latinos is early childhood education, and for African Americans it is social work. Ultimately, the country's emerging workforce needs will be met through self-responsibility as students and workers are exposed to incentives to learn, and community colleges will be a key infrastructure component in enabling them to execute on those choices.

## Lifting All Boats

And what about when things don't go as planned?

We've argued that the impacts of advancing technologies and guided migration flows that improve overall demographics are having a net-positive impact on the United States. And as we've described in this book, it is those at the margins of the economy—racial minorities, the least educated, the disabled—who actually benefit the most when an economy is growing quickly with full employment. So, from a social-equity perspective, the best first-order strategy for helping those at the bottom to improve their lives is overall economic growth.

But we also see how some people in society—or, rather, a broader swath of people at particular times in their lives—may nonetheless be hurt by these changes. What responsibility should society have to those who lose from the shifts of the twenty-first century, or who experience its shocks along the arc toward progress?

Technological changes and business response will always be uncertain, and no one can predict for sure who will be most impacted. In a recent *Foreign Affairs* essay, Walter Russell Mead described how it took decades for the technologies of the Industrial Revolution to mature and express themselves in society with clarity enough to give governance an effective response.[18] Beginning in the Reconstruction era and continuing through the early twentieth century, the US government and society gradually came to terms with the full benefits and costs of industrialization, a slow process of recognition and adaptation that culminated in the wartime mobilization during World War II. The Progressive Era and New Deal policy innovations, such as redistributive income taxation to support a social safety net, were hallmarks of that process. In some senses, we take that system of redistribution for granted today as a core role of US federal and state governments. But in fact it was a specific response to changing economic and social conditions of the time.

Contemplating the far side of today's hinge of history, we should prepare to muddle through these changes once again. We are probably not ready for sweeping federal policy reforms in the name of technological disruption. But we should be ready to think a fresh about both the true nature of employment today and, in turn, the goals of our existing safety nets—their sufficiency, and how to balance their costs and other trade-offs as conditions change.

What should we be looking for? Recall MIT's Erik Brynjolfsson's research on the suitability of some specific tasks within existing American jobs to being carried out by machine learning; it suggested a few possibilities. His first observation is that while all jobs are made up of at least some tasks that lend themselves to machine learning, no job is completely composed of tasks that make sense for machines to perform. So we should not expect entire labor sectors or even occupations to simply go away. Instead, we can expect an increased rate of business reorganization, reinvestment, or individual job redesign and reassignment as new technologies are adopted.

A few sectors are worth keeping a special watch on, given the share of their activities that are likely to be suitable for new investment in machine technologies such as image or voice recognition. These include accommodation and food services, transportation and warehousing, retail trade, and some types of manufacturing. According to Brynjolfsson, education and health care, by contrast, are two sectors that have less overlap with the currently emerging capabilities of machine learning or advanced manufacturing technologies; basically, while those disciplines may substantially benefit from changes, a smaller proportion of what people in those sectors spend their time doing today (e.g., interacting with other humans in person using emotional intelligence) can be done well by machines.

Within those target sectors, some workers are likely to become significantly more productive as they are able to use machine technologies to complement their existing work. Others may involuntarily find themselves working fewer hours each week as some, but not all, of their responsibilities get off-loaded, only for those hours to come back as businesses readjust. Or, we could see broad layoffs when individual firms fail to compete with in-sector rivals who thrive and expand market share by taking better advantage of these transitions. Brynjolfsson further suggests that people working in lower-wage jobs will see these sorts of changes most frequently, and that these shifts will be played out with relative frequency in smaller cities versus larger cities (where employment tends to be more knowledge based). In short, there is not a forecast for a broad decline in employment—in fact, there is likely to be overall gain, which is the whole idea—but the process could create churn in people's working lives. So there is a role for policy there.

Other analyses of the situation are less sanguine than Brynjolfsson's. While we cannot say with any certainty how much the emerging world will affect the labor force, the oft-cited, and more often misunderstood, 2013 study by Carl Benedikt Frey and Michael A. Osborne ignited a good deal of debate with its conclusion that 47 percent of American jobs were at "high risk" of automation.[19] Responding in part to Frey and Osborne, two working papers from the Organisation for Economic Co-operation and Development (OECD) analyzed the tasks performed in occupations to determine how many jobs could be automated and how the responsibilities and tasks performed by workers would change. One saw an average of 9 percent of jobs in OECD countries as "automatable"; though the number varied by country, the United States matched the OECD average in their analysis, with lower-income occupations more at risk than higher-income ones.[20] And the other study took the over, seeing 14 percent of OECD jobs as automatable and predicting that the tasks and nature of 32 percent of jobs would likely change significantly.[21] Researchers at McKinsey Global Institute (MGI) offered their own assessment in 2017, concluding that six of out ten workers could see 30 percent of their job automated by 2030. More useful for our purposes, MGI also studied the potential for job turnover as a result of automation and artificial intelligence; it concluded that 15 percent of workers could see their jobs displaced and 3 percent may find themselves changing job categories entirely.[22] Many more scholars have put their minds to this task of predicting the impact of new technologies on jobs, and their work echoes the range of predictions presented here. Consistent across these studies is a concern about the impact of job automation on workers at the lower end of the income distribution. We share this concern, as mentioned at the outset of this discussion, and recognize how destabilizing—on both a financial and personal level—job churn can be.

Importantly, job churn isn't new: the US economy already both creates and destroys more than six million jobs each and every month, with about two-thirds of those jobholders quitting and one-third being laid off or fired, with a net result of joblessness being essentially flat (notably, the US economy as of 2019 was actually estimated to have approximately 1.5 million net "excess jobs" still for the taking, given the right employer-employee match). And Bureau of Labor Statistics (BLS) longitudinal data indicates that American baby boomers held an average of 11.7 jobs and experienced 5.6 spells of unemployment during their working years from age 18 to age 48. Showing the importance of education, high school dropouts had more frequent unemployment, at 7.7 spells, while college graduates experienced 3.9 spells, about half the rate.[23] So mass job transitions are something our US labor market does literally every day. But we should nonetheless be thinking of ways we might actually wish to facilitate job transitions going forward so as to best realize the benefits of new technologies.

Back in the 1970s, Congress used to invite the relevant cabinet secretary to do bill markup in the chamber. A newly minted secretary of labor had come to Washington, fresh out of the comfort of his role as dean at the University of Chicago. One day, an unemployment bill was being discussed in the House, and he sat in the House Ways and Means Committee, reeling off matter-of-fact answers to the staff members on this and that. After a few rounds of this, the chair, Wilbur Mills, eventually turned and said to this executive branch interloper in a slow but sanguine Arkansas drawl, "Mr. Secretary, you might have gotten your way as a teacher in the classroom—but down here we *compromise.*" He then went around the room, member by member, Democrat and Republican, and each one of them had an idea of how some provision being described would affect their own district, and the constituents they represented—"Have you thought of X? What about Y?" Reality wasn't as clear-cut as the theory. And yes, Mills's political mind may have been thinking about pork. But there was more to it than sausage making. He was thinking about how one goes from a policy concept to a sort of shared, nonpartisan understanding. There is a human, individual nature to a job, the social safety net behind it, and how that might change in an emerging new world. So let's workshop two ideas, with that in mind.

## Encourage Work

The first policy principle to keep in mind in a changing America is to encourage work. Work is the only reliable path to prosperity. The US poverty rate—defined as an income less than three times what a family needs to buy food—is now just above 12 percent. But poverty in America is dynamic. According to statistics compiled by the Urban Institute in Washington, DC, the "average" American has about a one-in-twenty chance of falling into poverty in any single year, and just over half of Americans experience poverty at some point before they reach retirement age.[24] About half the time this poverty is caused by someone in the family losing a job. But it is also generally short-lived; about half of those who enter poverty exit it within one year, and three-quarters exit poverty within five years. And most of the time that exit is due to a family member's getting a raise or a new job. Some members of society face severe challenges that effectively prevent them from leading a prosperous life. But they are a small minority. In general, we should think of being poor as a symptom—not as membership in a class—and the diagnosis is generally one of being out of work.

Apart from prosperity, work also provides personal meaning. Psychological studies since the 1970s have shown the importance of structure in modern human happiness. And for most people, that structure comes from their job. Even controlling for loss of income, joblessness in this country and elsewhere around the world has been correlated with lower reported rates of well-being, along with depression, the loss of a sense of social belonging and motivation, and even physical health challenges. These feelings in turn result in the use of destructive coping mechanisms, such as substance abuse and family tensions. In short, we should want people to be working, even if they do not need the money.

In that sense, a safety "net" may be a misnomer. In our emerging new work environment, we may expect people to first experience layoffs at a younger age than is typical today. Capable, trained, and educated people may experience multiple layoffs or buyouts throughout their career. Job security will vary based upon the tasks one performs, and in nonobvious ways across sectors, so a broader swath of professionals may see themselves receiving pink slips. The flip side of this, of course, is that in increasingly dynamic sectors subject to reorganization and reinvestment, workers themselves may choose to jump ship more frequently for higher wages or more desirable opportunities in other roles or firms. And Americans will work longer as they live longer. In short, job disruptions of this sort may become normalized, expected even. Ideally, with the right individual responses, they could even come to be seen as positive, given that they force the issue of improving both overall and (hopefully) individual productivity.

Because of this, we should not think of US welfare programs as catching people as they fall and holding them there, just above the abyss. Rather, the country needs ways to get capable people back to work as soon as possible, or even to retrain them before they lose their jobs so as to improve their productivity at their existing jobs or firms. In a free economy, workers and employers themselves will always be the prime actors in this transition—think of those six million jobs lost and gained every month. In some cases, the government may be well suited to supplementing individual actions through its welfare programs, or in reducing the barriers it may have already put in place to finding new work, unintentionally or otherwise. Welfare of this sort should be recast as providing personal agency rather than reducing pain.

Given that the labor market realities that will occur in piecemeal fashion, across different sectors, and in different regions around our country, how does our current welfare and job-training system stack up in this regard?

Today's job-adjustment programs are fairly small; they are arguably ineffective; and most importantly, they tend to encourage behaviors that keep people out of work. Unemployment insurance (formally, the Federal-State Unemployment Insurance Program), for example, is a tangled web of federal- and state-level funding and administration. Established in the 1930s, the program sees states establishing unemployment insurance parameters and collecting premiums through payroll taxes; states later administer benefits (with the administrative costs of doing so actually covered by a separate small payroll tax established by the federal government, which also acts as a centralized treasurer for each state's funding pools). An interesting feature of this program is that employers who remit the tax through payroll have rates set based upon "experience": employers with a history of laying off workers must pay higher insurance rates, originally envisioned as an incentive for employers to retain employees over time. And because states can regularly adjust tax rates or benefit payouts based upon available trust funds, the fiscal health of these programs is generally good.

Laid-off workers generally can receive insurance benefits through their state for twenty-six weeks, but the federal government finances an additional thirteen weeks in some high-unemployment states, and during a recession beneficiaries receive additional weeks' benefits on top of that, in addition to other potential extenders, such as trade readjustment allowances. However, statistics show that a recipient's reemployment probability declines steadily from week ten to approximately week twenty-four, before spiking just before expiration at week twenty-six. This suggests that many recipients are in fact able to find new jobs but simply are not doing so in order to make full use of their allowed payments.

If being laid off is a rare thing in one's career, then it may sound acceptable to stay out of the labor force to regroup and recover for a full course of unemployment—half a year or even a year. Unemployment benefit amounts vary by state but often cover half of one's previous salary, up to some cap, so the "free" income could be welcome. In the long run, though, economists estimate that being out of the labor force for any period of time can actually reduce one's lifetime earnings by more than three times the direct lost income over that period, considering not only foregone wages but also missed raises (which compound over time) and missed retirement savings. And the younger you are, the more it hurts (the average age of unemployment insurance beneficiaries today is about thirty-eight). Unfortunately, even if workers are aware of this lost income dynamic, the appeal of unemployment benefits today versus long-term earnings in the future still presents a cognitive bias toward staying out of the workforce.

This issue will become more important. As described earlier, future Americans may experience higher rates of job churn on account of rapid technology changes (or perhaps from immigration surges in some parts of the country, depending on how well expected migration flows can be managed).

We expect that this higher rate of churn will result in higher productivity, and in turn better worker prosperity. If, however, we follow current unemployment practices, more churn could also mean more people with more accumulated time out of work over their careers. This moves us in the wrong direction on the prosperity front. We know that being out of work has long-term costs—both financial and in terms of personal well-being. So we need to be sure that the temporary financial cushion offered by unemployment insurance doesn't deter people from moving between jobs as quickly as possible, minimizing lost earnings and opportunity cost.

For the relatively narrow matter of unemployment insurance, one response to this incentive and behavioral problem has been policy proposals to let people take all twenty-six weeks of payments up front, then to use that funding to enroll in community college or other coursework right away to upgrade their skills. A change like that could be gamed by employers or employees in certain sectors—seasonal and agricultural work, for example—but with guardrails it could be a better approach than what we have today. Plus, there are a variety of separate long-term private and social goods to be had from workers being better educated. But this approach could also increase program costs if the state ends up raising overall payroll taxes to cover full courses of payments to everyone, when some people may have otherwise found work before their benefits were exhausted. A revenue-neutral fix to this could be to offer to pay out an amount equal to the average, rather than maximum, duration benefit (though this too could have other distributional impacts based upon class behavior). A truism in public policy: there are trade-offs.

To that end, an obvious reform, which has been used before, would be simply to reduce the period of benefits eligibility. Shortening the transitions lowers program costs, and, to the extent individuals are actually harmed over their lifetimes by being out of the labor force (even when they are receiving benefits), doing so might help beneficiaries too. Some states experimented with this in the past when their trust-fund balances ran low. It was also the basic argument underlying federal welfare reform and the 1996 Personal Responsibility and Work Opportunity Reconciliation Act. The result of two years of bipartisan congressional and White House negotiation in Washington, that bill focused on the Aid to Families with Dependent Children (now named Temporary Assistance for Needy Families) program, which was also created in the 1930s under the Social Security Act but had grown substantially beyond its original intent over the years. The reform was a fairly blunt instrument: by the mid-1990s, recipients, largely unmarried women with children, were on the rolls for an average of eight years, so it limited eligibility to five years. While not uncontroversial, the value of this change was one of information and signaling—young women saw that this would not be a lifetime welfare program. And it also changed the direction of government welfare offices from policing eligibility to helping people find work. We'll come back to this reform later.

However, the emerging world prompts us to consider another way of approaching the issue: should we look at unemployment as a catastrophic event that should be insured for, or should it be seen as a reasonably expected event, one that should be prepared for and saved toward in order to smooth out income and assets? To this end, rather than offering a fixed term of benefits upon a worker's being laid off, with funds pooled across an entire state's workers, we could instead change the concept to one of a personal unemployment benefit "savings account." As proposed by the late Nobel Prize–winning Chicago economist Gary Becker, in place of the current system, workers would accrue lifetime unemployment benefit credits as they work. These could be redeemed as cash when unemployed (or perhaps multiplied if used toward some investment like tuition expenses). Instead of the accrued benefits evaporating upon finding a new job, unused credits could simply be banked for the next layoff (or even made accessible to people who proactively perceive their own need to invest in reskilling education even while still working).

We might relate a personal story about a positive experience with a work-based educational credit accrual system administered by the federal government: the GI Bill. One of your authors gained deferred admission to an industrial economics PhD program at MIT after earning an undergraduate degree at Princeton in 1942. But the Second World War, which meant three years as a US Marine in the Pacific, intervened. Upon their return to Massachusetts, both he and his wife, herself a veteran and nurse, were able to access GI Bill voucher funds to launch new interests and careers (and even to guarantee their $10,000 mortgage on a house). The GI Bill was an example of the US military's facing the same sorts of questions that we expect will become more common among the future workforce: what do you do with capable persons who nonetheless quickly find themselves needing to take on new skills or entirely new careers in the face of changing labor needs? The GI Bill's answer was not to give veterans cash as a cushion against temporary poverty, or to start a new Washington-directed jobs program, but rather to explicitly invest in their educations or entrepreneurial motivations to enable self-reinvention and self-improvement as decided by the recipients themselves. Follow-ons from the GI Bill still help about 1 percent of Americans—veterans or their families—today. But it was and is expensive. It cannot be spread to cover the totality of the US population. Its principles, however, are worth mimicking: how to make oneself more job eligible and get back into productive work.

Unemployment insurance as we have described it here does not operate in a vacuum. Rather, it interacts through a web of other entitlements, job-training efforts, and childhood-support and poverty-alleviation transfer programs. Perhaps the best of these, from a work-incentives perspective, is the Earned Income Tax Credit, which was enacted in the mid-1970s and expanded in the 1986 tax reforms. The credit, which costs the federal government about $70 billion per year, is designed to magnify the earnings of prime-age, low-income workers (particularly those with children). For example, a married couple with one child living in their household will see an additional thirty-four cents for every dollar earned up to their first $10,000 or so in annual income. That credit is then gradually reduced toward zero as incomes rise beyond about $19,000, so it is well-targeted at those who need it most—working heads of household. The overall idea is to ensure that a parent who works full time will be able to keep their household out of poverty. So far, so good.

We might suggest a few changes, however, with an eye toward future American labor markets. Today's program tries to do many things: it is poverty alleviation, it is family oriented, and it recognizes the costs and needs of having children. Given that it also incentivizes work—recipients must earn income to get the credit, hence the name—it might be valuable to apply that compulsory function to a broader swath of the population. For example, we know that the American workforce is aging and the population is living longer. As these demographics shift, we also need to expand our outdated notions of what "working-age population" means. The Earned Income Tax Credit currently phases out for workers at the age of sixty-five, presumably with the idea in mind that recipients of that age will have Social Security to fall back on, or perhaps that they are not supporting children so they no longer need the credit. And in a reversal since the enactment of Social Security, elderly Americans as a group are also now among the wealthiest of American age groups, which argues against further subsidy. But there is still personal and social value in work, and this population is one that we should be thinking about incentivizing to stay on the job. (Hoover economist John Shoven has recommended achieving a similar goal by simply removing existing Social Security payroll taxes for workers over sixty-five, to increase their after-tax incomes from working.) Some states have begun to enact their own versions of the Earned Income Tax Credit to supplement the federal core, for example by expanding coverage to groups such as the elderly or single young workers. California has done this. So has Minnesota.

Fraud, unintentional or not, is a major problem with the program, with the Government Accounting Office estimating that nearly a quarter of recipients receiving the benefit should not in fact qualify, so this is something that states should be on the lookout for. And of course it quickly becomes expensive to expand eligible recipients too far up the income scale when trying to cover other groups of workers. Instituting new payroll or income taxes (which discourage work across the population broadly by reducing take-home income) in order to expand a tax incentive intended to encourage work can get out of hand, so, again, revenue-neutral reforms would be ideal. But, as different states perceive different needs, this is one tool that states should keep in mind as they watch their own labor markets transform in the years to come.

If the Earned Income Tax Credit is an encouraging example of welfare programs that affect work decisions, let's also turn to where such programs can go wrong. Consider the situation today when a fifty-year-old small-town warehouse worker gets laid off. She collects unemployment. But she also applies for and receives Social Security disability insurance for chronic pain, based upon broadly expanded health eligibility standards. The local economy seems uncertain. She considers moving for an appealing paralegal job offer near her sister, two states away, but doing so would require a new state certification and she doesn't want to leave an aging mother who needs home care. The loss of health insurance from being out of work is daunting, but she learns that with the disability designation, she could become Medicare eligible within two years. Taking these well-meaning programs all together, she sees a huge incentive to simply stay out of the labor force altogether, or she risks losing substantial benefits.

In fact, recent research from Harvard economists Nathaniel Hendren and Ben Sprung-Keyser estimates that spending on these later-in-life safety-net programs—adult health, disability, and unemployment insurance itself—have among the lowest returns on investment for the federal government (in terms of lifetime tax receipts), given that many of them end up encouraging people to stay out of the workforce. Moreover, beneficiaries of one welfare program often end up enrolling in another as well, further compounding government spending losses. Rolls for disability benefits, for example, grew from 2.7 million in the mid-1980s to 9.0 million in 2014 alongside an aging population and a widening definition of eligibility, despite workers on average reporting feeling healthier and jobs becoming less physically demanding. Enrollment in multiple welfare programs also further reduces personal incentives to find work—at some parts of the income scale, the "effective marginal tax rate" on increasing one's income can actually exceed 50 percent given the concurrent loss in welfare benefits from doing so, such as monthly payments and Medicaid. (As an extreme example of the phenomenon, many American workers covered under the 2020 CARES Act, passed quickly in response to COVID-19 economic shutdowns, actually faced effective marginal tax rates of returning to work of over 100 percent.) Reforms to more strictly define beneficiaries as those who are unable to do work of any kind—as opposed to those unable to perform their existing profession—will be important as the entire US population shifts toward more dynamic employment environments.

Overall, it's clear that our safety-net programs need some sort of a balance—not just in how much money the public collectively spends on them, but also in how incentives to work are presented to the recipients.

## What Level of Government Will Be Most Effective?

This brings us to our second policy principle, which is one of policy diversity.

We can think back to Chairman Mills's markup committee admonishment, and to Walter Russell Mead's read on the history of technology and society. Today we can only broadly speculate on how future labor market disruptions will or will not play out. So at this point we can give one clear negative policy recommendation: the federal government should not attempt to put its thumb on the scale at this early stage with a sweeping policy announcement until the market has had a chance to digest the changes now emerging. States, however, should keep their ears and eyes open, and be ready to get to work. States and localities are closer to the actualities of the job situations—challenges and opportunities, sector specifics, educational and retraining capacities, options for relocation—that their residents face. The federal government, on the other hand, will always have a degree of inflexibility in managing these sorts of programs from on high. So states should be encouraged to experiment and learn from one another as something works, or doesn't work. They should lead.

The 1996 welfare reforms came about by way of just such experimentation. Toward the end of President Reagan's administration, Congress passed a bill that allowed individual states to experiment with alternative reforms to the Aid to Families with Dependent Children program described earlier, through federally issued waivers. For example, states could stipulate that program benefits be used for education or for childcare, or they could limit time on eligibility rolls. Though the number of welfare recipients continued to grow rapidly, from 10.7 million recipients in 1988 to 14 million five years later, for these first few years that waiver program went largely ignored at the federal level. Having campaigned on the issue, though, President Clinton entered office and started granting reform waivers to a number of states to reduce the rolls. Over the next few years, results of these state-level reforms started coming in, with both Democrats and Republicans liking what they saw. By the time national legislation was being debated, early experimentation had demonstrated that states were capable of "doing the job," and their effectiveness gave confidence that similar reforms could be done at the national level. The number of mothers formerly reliant on the program has since fallen by two-thirds (they have entered into the workforce), and US poverty rates have continued their decline.

States aren't always perfect. Welfare programs in particular demand scrutiny, and the reason why so many of these programs were federalized in the first place was on account of states, particularly in the South, that discriminated against their black populations through the guise of welfare eligibility rules. So reforms necessitate care. But with that care, we are hopeful that states could once again offer some creative early efforts at tweaking the United States' approach to unemployment, and reemployment, more broadly. Some may get it wrong, but others will get it right, and others still can then learn from what works. The emerging new labor market will be defined by faster-moving economies with shorter job tenures; an older and more experienced workforce with potentially fewer children at home (though worker age and family size will vary by race and class and national origin); a shifting idea of work itself, with side gigs that are enabled by technology and workers not getting laid off so much as continually readjusting to different hours; the need to move new workers from less productive to more productive areas within states and around the country—or, indeed, move them into working remotely; and, if all goes well, a stronger overall economy, with higher productivity and raw growth rates (and more prosperous citizens) to offset the need for such programs in first place. There is surely a variety of solutions to this complex changing reality, but they are not likely to emerge fully formed from an enlightened congressional committee meeting on a summer afternoon beside the Potomac.

Focusing on a particular location—a city or town, rather than a broad class of people—can also be a way to capture the impacts of a changing economy. We described how much individuals move into and out of employment situations or income brackets across their lifetimes. This makes it difficult for even a "means-based" social safety net to identify who truly needs help, and who doesn't. Places, however, carry some momentum. Geographic inequality is a palpable thing. It is obvious to see when a town is in decline, or when it is on the rise, and how its atmosphere filters into residents' job choices, opportunities for personal improvement, and quality of life. Even health is impacted. The federal government should not take responsibility for preserving every aging metropolis—allowing the process of disruption, balanced by growth elsewhere, is part of harvesting the gains of technological change—but it should consider the framing of geographies alongside individuals in defining its safety nets. Data can also help, for example by monitoring and reporting the vitality of American cities and towns, their job markets, and the quality of their educational opportunities in such a way that ordinary Americans can see what is going on around them and what choices they may have elsewhere.

Finally, the firms themselves have a role to play. It is the government's responsibility to cultivate a social foundation that serves the overall good. This means education, public safety, infrastructure, information, welfare where needed, and so forth. But it is not the government's sole responsibility to deliver an optimized labor market for the profit of passive employers and their shareholders. As MIT's Brynjolfsson argued in his analyses for this project, while advances in technology are often breathtaking, reorganization of US businesses has actually slowed in recent decades, and the reskilling of the workforce has lagged: "The key bottleneck for unlocking value often is not technology, but people." Our Hoover colleague economist Michael Boskin is fond of pointing out that the federal government alone now runs forty-six job-training programs spread out over nine agencies, costing over $20 billion every year, with essentially no metric for success.

And actually, the United States has been here before. In the late 1950s the American public was seized with fear of automation in the factory—that blue-collar jobs would be destroyed by the introduction of industrial robots. One of us even wrote an essay on the topic in 1955 while an assistant professor of economics at MIT to offer some implications of the change for factory managers.[25] Nonprofit foundations took up the topic. Congressional hearings were held. And the federal government decided it needed to be involved, as well. The result was President Kennedy's Manpower Development and Training Act of 1962, which got the federal government directly involved in classroom and on-the-job training for head-of-household earners. It started well but over time grew to be less useful for the changing labor and social environment. And as it turned out, the job growth rate in the 1960s was the highest in decades—maybe no program was needed at all.

Despite this, by the time of President Nixon's administration, there was a new effort, the 1973 Comprehensive Employment and Training Act, which focused on providing short-term public or nonprofit jobs to youth. Following reports of waste, that too was replaced during the Reagan administration by the 1982 Job Training Partnership Act, which later became President Clinton's Workforce Investment Act of 1998. In each of these efforts, the markets moved faster than the decadal-scale federal legislation packages. Hoover's John Cogan, for example, recounts his experience as assistant secretary for policy at the Department of Labor in the early 1980s. One of the strategies of the Nixon-era jobs program had been to try to forecast the labor needs and disruptions of the coming decade in order to guide its own training efforts. But the department's retrospective analysis of its macro-level predictions later showed that original guidance (and therefore the training efforts linked to them as well) to have missed the mark: the advent of the computer revolution, for example, had been completely missed.

Moreover, all such efforts from Washington have been hobbled by the same problem: despite the best of intentions, or the best efforts by those who run or participate in them, the federal government does not have the direct incentive or information needed to be sure that such programs are effective and useful for private enterprise. Firms do. Moreover, as Brynjolfsson's research has shown, an employee's skills have not just personal (transferable) value but also firm-specific value.

Is there a way, then, to encourage firms to further invest in worker skills, in effect making them an on-the-ground extension of the public social safety net? Firms have the interests and perhaps the best information about training employees for changing workforce needs. And, as a business cost, employee training expenses are already "tax deductible" from an accounting perspective. But many US businesses are not experienced in education or formal training. Nearly six hundred thousand Americans do participate in twenty-four thousand different federally certified apprenticeship programs each year, but their reach is limited; they are dominated by union training programs in the construction industry (and by the US military). For the rest, it will be important to reduce the transaction costs that businesses face in training or retraining employees or potential employees.

As described earlier, Van Ton-Quinlivan took on this very subject in her time overseeing California's sprawling community college network. While these institutions are excellent assets overall, she found that given the number and unevenness in quality of small two-year colleges—and funding limitations—they were not making sufficient progress in coordinating their workforce-training curricula with private businesses. The return on effort for local firms in working with individual colleges was simply not worth those firms' time. But by using state funding to incentivize small groups of colleges to form regional curricula and teaching consortia, the results of a private firm's collaboration could then be magnified across a broader pool of students, interns, and potential future recruits. New partners signed up, and workers benefited. Every state may face a different barrier to this sort of community college–employer collaboration, but the broader point is that there is already in place an existing educational and training resource whose capabilities and needs are a good complement to private firms. Earlier we noted expectations about the sectors that may see the highest uptake of machine learning or robotics technologies given the applicability of those capabilities to their workers' current responsibilities: retail, food service, warehousing and transportation, and some types of manufacturing. Were state or federal governments to dedicate new spending or tax benefits toward such collaborations, these sectors could be good candidates.

For more:

- "How Will Machine Learning Transform the Labor Market?" by Erik Brynjolfsson, Daniel Rock, and Prasanna Tambe
- "Pathways to Economic Opportunity in the Twenty-First Century: A Case Study on How the California Community Colleges Modernized to Deliver on Its Workforce Mission" by Van Ton-Quinlivan

*https://www.hoover.org/publications/governance-emerging-new-world/spring-series-issue-619*

## From "How Will Machine Learning Transform the Labor Market?"

*by Erik Brynjolfsson, Daniel Rock, and Prasanna Tambe*

The branch of AI known as machine learning (ML) has advanced significantly in just the past decade, largely reflecting improvements in the area of deep learning, a technique that trains large neural networks on large data sets. Three different types of advances, each of about two orders of magnitude, have combined to make this possible: 1) an increase in the quantity and quality of digital data; 2) improvements in computational power, reflecting not only the march of Moore's law but also new specialized architectures like GPUs (graphics processing units) and TPUs (tensor processing units); and 3) improved algorithms. As a result, the performance of ML algorithms has improved significantly. In a highly cited example, the image-recognition algorithms on the ImageNet Dataset improved from barely 70 percent in 2010 to over 97 percent today and now surpass human-level performance on the same data. Voice-recognition and natural language processing, machine translation, recommendation systems, gaming, and many other tasks have also seen striking improvements. Because capabilities like vision, speech, and decision making are so fundamental for most occupations, these improvements to technology suggest that substantial changes in the nature of work can be expected.

Despite these impressive advances, however, ML is far from being capable of doing the full range of human cognitive tasks. This raises some obvious questions. What tasks can ML do well, and what tasks are best done by humans?

# CHAPTER 8

# Emerging Technology and America's National Security

For all the throes of our economy or politics or social advancement that consume us day to day, these are as but a play staged atop a thin shell. And each day, that fragile stage is patched and refinished, through great shared cost and personal sacrifice that create a global security commons. It, too, will face new challenges across the hinge of history that must be successfully navigated in order to act upon other currents we have described in this book.

When looking at that global security environment, we are reminded of President Reagan's approach to dealing with a complex and dangerous world. The first order of business, he felt, was to be realistic about the world around you. Then you had to be strong in all senses of the term—militarily, economically, politically, and in national spirit. Finally, as you went out into the world, you had to set your objectives—know what you want—and focus on that agenda. It was a wise, and ultimately successful, approach.

Today we see great challenges arising in the global security realm but also great opportunities to create a safer, more secure world. We see the emergence of new technologies and the development of new ways of utilizing them for military means. China is adopting these new tools particularly well and devising effective concepts and strategies for their use; Russia is employing internet and communications technologies while also developing certain high-end capabilities; and nonstate actors are gaining access to new, increasingly lethal weapons.

More broadly, we recognize that the advent of new economic centers leads to new technological centers, as in China. And we see the globalization of technology: new and emerging technologies are being developed across borders and across disciplines. The nature of these technologies contributes to their globalization. Some, such as additive manufacturing, democratize production, while effective and creative applications of artificial intelligence (AI) are being fostered openly and internationally. We need to understand how these phenomena are changing the face of international security and what they mean for the United States.

At the same time, we can make sure that the United States continues to operate from a position of strength. Fortunately, there's a good foundation. The United States has led the modern world in science, technology, and innovation, and that leadership underpins American economic and military supremacy. The United States' scientists, mathematicians, and engineers excel at fundamental research but also at transforming that research into usable technology, realizing new military capabilities. This nation owes that advantage to any number of factors, among them an entrepreneurial spirit and culture of innovation, centers of scientific excellence at universities, a conducive business environment, and a productive relationship between the public and private sectors. And the United States also enjoys an expansive network of partners and allies who excel in this arena.

Although challenges to US preeminence have appeared and are chipping away at our technological edge, the rise of new technologies offers great opportunities for the United States. Revitalizing the national tradition of excellence in the development of critical technologies and their practical applications will contribute to both the nation's national-security objectives and our economic prosperity—and to our ability to lead international-security efforts.

The rise of new technologies will not, in itself, invariably shape the future. If we recognize the challenges and opportunities before us, we can develop sound strategies to strengthen our own innovation base; the United States can work with allies and partners but also with Russia, China, and other nations to shape a more stable future.

## China and the Indo-Pacific

The United States and the People's Republic of China (PRC) share extensive ties in trade, investment, science, and diplomacy, and citizens of the two countries maintain historically deep personal connections. Overall, this is a relationship with strong mutual benefit. At the same time, the United States engages in competition with China on economic, technological, military, and even ideological fronts. The PRC's integrated information-technology strategy, in particular, makes it uniquely capable of disrupting the liberal order championed by the United States and its allies across the broader Indo-Pacific region, which stretches from the Indian Ocean to Hawaii.

Importantly, the nature of these fields of competition is new. Kinetic engagements—what we generally think of when we think of war—are of course central to any military conflict, but, as Admiral Gary Roughead (USN, ret.) and the Applied Physics Laboratory's Emelia Spencer Probasco and Ralph Semmel argued in our discussions of this subject, information itself is transforming both the nature of competition and what it means to win in war. The United States' technological innovation and cultural attractiveness has in recent years allowed it to enjoy information dominance, and now the Chinese Communist Party (CCP) has made a national commitment to achieve that same objective.

It is easy to speak about the opportunities presented by technology but another thing to realize them. In the military sphere, China has taken that latter step and embarked on a national effort to take advantage of information technology and control. It is building institutions to coordinate the fielding and use of high-tech capabilities, such as its Strategic Support Force, which integrates cyber, space, and electronic warfare operations. It has leveraged its status as an emergent and expanding power to invest in leap-over technologies that potentially skip entire generations of built-up know-how and lethality. And the party-dominated state mandates civil-military cooperation, forcing a whole-of-nation approach and benefiting from its innovative and entrepreneurial society.

The latter point deserves some attention. We have already described how Chinese innovation is not solely state directed but organic and extensive. China has significant risk tolerance, which encourages the rapid and widespread adoption of new technologies. Moreover, US corporations—consider Alphabet/Google—seek to reach global markets, while their Chinese counterparts may have a different set of incentives or trade-offs. The CCP has established legal authority to demand cooperation from private entities.

The CCP and the People's Liberation Army (PLA) have succeeded in the "gray zone"—broad approaches to competition below the threshold of shooting conflict. Where US policy makers often think about gray-zone competition in geographic terms—China's use of "coast guard" or fishing vessels rather than military craft to skirmish for territorial control vis-a-vis neighboring countries in the South China Sea, to pick a common example—Chinese authorities actually implement such strategies in a deeper and cross-functional way, overlaying those noncombat surface maneuvers with physical infrastructure investments, coordinated information campaigns, legal actions, coercion, and other applications of global influence and power. The fruits of their labors can be seen in the aforementioned South China Sea, where some believe China has executed a fait accompli.

New information infrastructure and applications put China in a position to extend its influence in telecommunications, information systems, and e-commerce in developing countries and other strategically significant locations (including outer space). China's innovations could end up serving as the foundation of the future information society and position China for technological and political advantage should its standards, systems, and policies be adopted widely by others. Its efforts to establish a global foothold in this respect have met some resistance, as can be seen in the ongoing disputes over whether the Chinese telecommunications company Huawei should or should not be allowed to supply new 5G networks in the United Kingdom and other US allies' territory.

All told, China employs a comprehensive, national approach to developing, fielding, and employing information technologies, many of which are dual use—valuable for both civilian and military enterprises. At the same time, it faces some significant structural problems of its own, which may hinder its ability to meet its own expectations: a poor demographic outlook; a slowing economy that is weaker than advertised; and an authoritarian government that contrasts with the successful governance alternatives presented in other Chinese societies like Hong Kong or Taiwan. Moreover, US private organizations still lead in high-tech development, though Chinese firms are close behind and getting closer. What distinguishes the Chinese approach to new technologies—and what deserves further discussion—is the way the PLA is integrating them across all domains of warfare, from undersea to space.

In their contributions to this project, Roughead, Probasco, and Semmel considered what might happen if President Xi tried to realize the long-standing CCP goal of taking control of Taiwan. China's investments in information technologies and PLA capabilities, coupled with sophisticated operational concepts, would give it a wealth of tools for forcing the issue. It could put on an impressive show of force, deploying submarines and other naval assets into the Taiwan Strait while test-firing missiles—two well-funded and developed capabilities. At the same time, it could activate an undersea sensor network using unmanned undersea vehicles; disrupt the American Global Positioning System (GPS) without interrupting the parallel Chinese BeiDou system; conduct a social media–based weaponized information campaign to undermine political will and confidence in the United States; interrupt power supplies in Taiwan and neighboring islands; control global shipping in and out of ports it owns; and so on. The Taiwan scenario is a compelling one, and one that shows the potential of well-integrated technologies and creative strategy and operational plans. And it highlights the importance of specific, high-end technologies to the future of conflict in the Indo-Pacific region:

*Artificial Intelligence:* For both sides, an important political precondition for conflict is that it would be "quick, decisive, and ultimately deflating to the adversary." Roughead and his colleagues posited that AI and the autonomous systems it enables "could affect the entirety of the information life cycle—how we collect it, secure it, manipulate it, defend it, share it, process it, integrate it, and act with it." Artificial intelligence has the potential to speed up the pace of conflicts, both forcing and enabling quicker decision making and responses. At the same time, it can facilitate coordinated, multidomain operations—both offensive and defensive—and allow control and manipulation of intelligence and information, as with deepfake videos. And AI also enables drone swarms and other advanced autonomous systems.

The history of AI development shows intermittent "AI springs," during which researchers make meaningful advances for a short period of time before progress plateaus yet again. What we are seeing today may be a fundamentally different scenario: the development of new applications of AI for civil and military purposes. China excels at AI research and applications, generating the most AI-related patents—a crude measure, to be sure, but a telling one—and has set up legal and political mechanisms to ensure the PLA and government entities have access to privately developed technologies.

*Cyber:* Although not an "emerging" technology, cybertech is fundamental to information warfare. It is ubiquitous but vulnerable, so, to quote the Defense Science Board, "defense is a necessary foundation for offense." That means defending military platforms and networks during operations but also protecting intellectual property, supply chains, military networks generally, and personnel information. This takes personnel trained in computing disciplines, and China has an impressive supply of experts coming through the university pipeline. Good defense will become harder for everyone with the arrival of 5G networks, which will create high-visibility targets for the collection, manipulation, and sharing of information. As referenced above, China aspires to develop and define the standards for 5G and to own the infrastructure. If it succeeds, it could wield decisive influence across Eurasia; the feasibility of auditing the security of a communications supply chain is unclear.

*Space:* Both the US and Chinese militaries rely on space-based assets for information transfer and positioning, navigation, and timing information. Space is central to military operations, to the effective use of precision weapons, and greater still to the functioning of the global economy.

Our space assets were designed for an uncontested environment, which we no longer have. Launch capabilities through competitive public-private partnerships, including for military assets, are an emerging bright spot in US technological capabilities. Nonetheless, last year, China surpassed all other nations in orbital launches, and it has demonstrated significant antisatellite capabilities. The US government remains unnecessarily tight-lipped about the challenge here. It ought to share more information about it, be more vocal about the challenge, and advocate for what must be done to assure access to such a vital domain.

*Others:* Other key issues include quantum technologies, and electronic warfare (EW) and directed energy (including high-powered microwave or laser weapons). The first of these qualifies as a true emerging technology—one still in the preliminary stage of development. Though it seems perpetually "twenty years away," quantum computing and sensing would accelerate AI, revolutionize sensors, and render public key encryption obsolete while also introducing incredible opportunities for positive technological advances. EW and directed energy are more narrowly confined to military use than other high-tech tools listed here, but they fall alongside cybertech in the continuum of information technologies. Directed-energy weapons, for example, may be key to defense against autonomous systems. Both the United States and China are working hard in each of these areas, with China arguably leading the way in EW.

## Recommendations: China and the Indo-Pacific

China has undertaken a long-term military modernization and reform program designed to prepare it for information-based competition. Its navy outnumbers that of the United States and its allies in the Western Pacific, and the PLA has planned carefully for a potential conflict with us. The United States, as the National Defense Strategy Commission warned, would struggle to win or would maybe even lose a war with the PRC.

But, again, China has significant internal demographic, economic, and political challenges of its own, and this competition is a national one, not merely a military or technological race. Chinese leadership will have to answer some difficult questions about whether it can sustain the current rate of military spending growth and continue to bring new opportunities to its people. Former US secretary of defense Jim Mattis stresses the strategic link between a country's economic "solvency" and its "security." Consider a history of three Chinas. The first modern China was that of Mao Zedong, a leader who brought it together, but in a dictatorial and cruel way that ended up limiting himself and the Chinese people. Then came Deng Xiaoping, who opened the country internally to the take advantage of the great strengths of the Chinese people, but who also had the winds at his back of a growing labor force (GDP growth equals a country's productivity growth plus its workforce growth). With this demographic dividend, China soared. Then, however, came the third China, Xi Jinping's China: its demographic dividend is used up, in part due to Deng's one-child policy, and while there is still some potential for further movement from rural to urban areas, the cream has already been taken off the top. So China's outlook for economic growth is spotty.

From that perspective, the United States is well positioned. It remains the preeminent power—economically, militarily, and technologically—and America's liberal system opens us to vast amounts of human potential. The challenge—one, for example, being revealed to us in the unfolding of the COVID-19 pandemic crisis—will be to muster the great national power and competence.

If we narrow our gaze to new technologies, as we have here, we can identify specific steps to take here at home to do just that. To begin with, the United States should focus more on the speed of practical applications than on revolutionary technologies. That is, think about how best to get new technologies out of the laboratory and into the field quickly and then use them most effectively.

Productive civil-military integration will be key to that effort. As discussed above, the Chinese system mandates effective exchange of technologies and concepts, but the US relationships among government, academia, and industry, which have traditionally powered American innovation, could be stronger still. The national-security and defense strategies label this ecosystem as "the national-security innovation base" and call for the strengthening and protection of that base; we concur. More broadly, the United States should focus on bringing public leaders and technologists together, at all ages.

From a strategic perspective, we can support the development of even more intellectual capital. Over 13 percent of American adults hold advanced degrees, a share that has steadily risen from 9 percent two decades ago. But American students now constitute a surprisingly small portion of US science, technology, engineering, and math (STEM) graduate programs, and the government struggles to train and retain high-quality civilian talent. The immigration system should be improved to attract and keep talented people in the United States, and the government can do more to encourage the intellectual development of young experts. Beyond US shores, the nation's allies and partners possess great intellectual capital of their own, multiplying our collective capacity. It will be important to maintain and strengthen those relationships and to build public support for them both at home and abroad.

Of course, China's population outnumbers America's, and China enjoys great human, intellectual capital of its own. It is incumbent upon the United States to recognize reality, to educate the public about this new reality, and to be clearer and more analytical about national strategies—for example, reconsidering the approach to gray-zone competition. As technologies increase the speed of decision making and warfare, we should also remain mindful that speed may help on the battlefield, but in strategy it can lead to instability.

The United States enjoys important advantages across the spectrum of technologies and high-end military capabilities and from its network of allies and partners, as well as its own strong, open economy and society. We should strengthen and sustain those pillars of national strength while moving the relationship with China beyond a zero-sum competition. Operating from a position of strength and confidence, the United States can work with the Chinese to build a healthier, more productive relationship. There is room for much engagement, even alongside our substantial disagreements. If we ever had to agree with the Soviets on everything before we could work with them on nuclear weapons, the Cold War would still be on. We can work with China, and think about constructive things to be doing with them today.

## The European Theater

Turning our eyes across Eurasia, we see a revisionist Russia and a NATO alliance in need of greater political unity. Russia, faced with significant strategic disadvantages of its own—among them poor demographics and a weak economy—knows it cannot match the United States and its European allies across the board. Instead it looks for those areas in which it can compete; in the words of one participant, it seeks "multiple levers against the West." General Philip Breedlove (USAF, ret.) and Georgia Tech professor Margaret Kosal argued, as we have already outlined in earlier chapters, that Russian development of high-end technologies should not be our primary concern. For all of President Putin's rhetoric about the importance of AI, his government has done little to foster a true innovation base.

Russia's "levers against the West" do include some emerging technologies, such as hypersonics and autonomous systems, and we should not lose sight of its development of those capabilities. But for the most part, it focuses on information warfare, certain asymmetric capabilities—such as integrated air defenses and long-range artillery—and nuclear weapons. Arguably, Russia has done the most damage to the West through its cyber-enabled political and information warfare campaigns.

Russia seeks coercive power through information manipulation and control. It exploits political divisions in Europe and the United States to weaken NATO and undermine confidence in Western, democratic systems. The 2016 presidential election may be the most obvious example, but Russia interferes in elections and political processes across Europe as well. Though President Putin's efforts seem opportunistic—targeting divisions or weaknesses as he sees them—the objective is clear: Russia seeks to sow discord and confusion, thereby imposing long-term, significant costs on the West, especially the United States.

To achieve that end, the Kremlin has closed the gap between the military and the rest of government, integrating nonmilitary capabilities into military operations to conduct full-spectrum competition—a type of "whole-of-government" approach. It also leverages relevant technologies to support that effort: realizing the discordant potential of social media on the low end and the value of autonomous systems at the high end, for example. And its operational concepts are innovative and deadly. Its proxies have used drones to coordinate and target artillery barrages rapidly and to great effect, proving that the ability to employ technology to generate strikes can sometimes trump the size of battalions.

The contrast between the Russian whole-of-government approach and the limited US response is striking. We are not the first to say the United States can do a better job of mobilizing all aspects of national power to the challenge, but it is true. Despite America's economic, military, and political advantages—and our close allies in Europe—we have not put up a good stop sign for President Putin. We have relied on economic measures to punish Russia, ignoring the array of other tools at our disposal.

## Recommendations: The European Theater

How can the United States bring its vast capacities to bear in the European theater? To begin with, we should maintain our technological lead. Though Russia is not a technological powerhouse, we can achieve more asymmetries through high-tech capabilities. That effort need not be undertaken alone; the United States should encourage technology transfer and cross-border development with its allies and partners. Sweden's cooperation with NATO in this arena is a good example of broad-scope tech development. Of course, as we have seen before, the United States must be careful to protect technologies as we go. Bad actors will often target smaller contractors and less sturdy governments—the weak links in the supply chain—so careful civil-military integration throughout the NATO alliance and its partners will be crucial. But we know that can be done; we can look to Estonia, which, in response to debilitating "test-run" Russian cyberattacks aimed at social intimidation in the spring of 2007, mobilized across government ministries and sectors—including banking, media, and electric utilities—to see how a nation can harden its infrastructure and institutions and master its own destiny.

Taking a broader view, we return to the oft-referenced idea of a more balanced, whole-of-nation approach. It bears repeating that the West can do a better job in the information arena and telling the story of US and NATO values and how we operate. We can truthfully promote our ideas and call out bad actors, and we can be less linear and formulaic in our behavior, employing information operations but also simultaneous diplomatic efforts to get off the defensive.

Of course, such a united effort requires political will and coordination, both at home and throughout the alliance. For four years running, NATO members have increased their defense spending. They have, as mentioned, bolstered their efforts to cooperate on cyber issues and elements of information warfare. And they have shown firm resolve against Russian aggression, as in the deployments of battle groups to the Baltic states and Poland. However, the political cohesion of the alliance is not as strong as it could be. If NATO members come closer together again, the alliance will be in even better shape to take on these challenges. As one participant noted: if we cannot build a political narrative of our own, shame on us.

Recall the 1980s, when the leaders of the United States and NATO agreed to deploy Pershing II nuclear missiles in Europe as a response to the Soviet Union's own nuclear deployments and saber rattling. It required a massive diplomatic effort and a healthy, united alliance—and President Reagan's open appeals to the hearts and minds of a skeptical European public to reinforce their own leaders' support—but it turned the tide of the Cold War. It was a nonlinear response to the Soviets, an unexpected but calculated escalation that changed the facts on the ground, a stop sign. We need another stop sign today. A new Pershing moment will likely not be based around nuclear weapons, but it will require renewed American leadership and a revived NATO.

## Nonstate Actors

Finally, let us turn our attention to nonstate actors. Though discussions of how the United States should deal with nonstate actors often fall to the tactical or operational levels, National Defense University's T. X. Hammes argued that we ought to look at better strategies. Insurgents, terrorists, and criminals are adopting new, but not cutting-edge, technologies and employing them in innovative ways, including AI-enabled autonomous systems. As those technologies become more accessible, they will give nonstate actors the kind of affordable, long-range weapons major powers have generally had to themselves. Counterinsurgency and counterterror operations will become increasingly challenging. American planners will have to change how they think about intervening abroad.

Insurgents generally prefer to employ available and widely used technologies. During the Iraq War, they used such commonplace items as garage openers and then cell phones as detonators for improvised explosive devices. Now, in Syria, we see increasing use of unmanned aircraft—commercial drones. Autonomous aircraft are quickly becoming cheaper and more capable while task-specific artificial intelligence improves their operations and multiplies their uses. A drone equipped with a camera, for example, can employ high-quality facial- or target-recognition software—consumer technologies already available to hobbyists—to become a targeted weapon. At the same time, additive manufacturing (3-D printing) makes them easier to build or repair in the absence of a dedicated supply chain and may, in the near future, allow inclined parties to mass-produce them.

Hammes reviews the various ways in which insurgents or terrorists might use autonomous systems. Coupled with the sort of easy-to-produce armor-piercing explosively formed penetrators (EFPs) pioneered in Iraqi insurgency "coffee-can" roadside bombings, they could target vehicles, disabling or even destroying them. A simple thermite grenade dropped onto fuel or ammunition dumps could ignite a conflagration, as has happened in Ukraine. Drone attacks on a civilian airfield could disrupt air travel, while one on a military airstrip—or a resupply depot, or a convoy—could interrupt logistics.

In sum, these new technologies will allow insurgents and terrorists to target and hunt specific targets. At the most basic level, troops in Iraq and Afghanistan have spent the past decade and a half staring at their feet for hidden improvised explosive devices (IEDs); now the IED can come to them.

Other advantages accrue to insurgents. Drones are no longer dependent on GPS; they now rely on inertial and visual navigation, and nonstate actors can locate targets through cheap space access—namely Google Maps and Google Earth. The ability to attack specific targets makes physical infrastructure and public figures more vulnerable, favoring the disaggregated insurgent force and working against the established power.

These technologies exacerbate existing obstacles to effective counterinsurgency operations, but the effectiveness of those efforts has always depended on strategy. As Hammes puts it, nations lose strategically, not tactically. When we commit to nation building, we commit to a long-term conflict and a heavy footprint. But as insurgents arm themselves with these new capabilities, they can further exploit that large footprint, targeting US bases, political infrastructure, and the like; it becomes even more challenging and costly to maintain a presence and establish general security.

The logical response to an enemy holding static or large formations at risk is to disperse and minimize your footprint. It would seem, then, the best response to a newly capable insurgency would be first to avoid large, direct interventions in the first place and to harden facilities—overhead protection, for example—as needed. Of course, the former decision is a fundamentally strategic one. At minimum, policy makers and strategists must be attuned to technological changes and adapt their understanding of counterinsurgencies accordingly; we must avoid faulty assumptions in a rapidly changing dynamic.

We return to President Reagan's approach to a changing, complex world. What does the emergence of new technologies mean for international security? How can the United States keep itself in a position of strength? And how can it help stabilize the international order and set the conditions for a more peaceful world?

We recognize that Russia, China, and nonstate actors present fundamentally different challenges to the United States, but for each we must deal with the advent of new capabilities from a strategic perspective. As nonstate actors gain access to increasingly capable drones, for example, US military strategists will have to rethink their approach to fighting them and engaging with broken states. Russia, meanwhile, presents a challenge in specific areas, including nuclear weapons, high-end offensive cyber capabilities, and the low-end technologies of information warfare. The United States and its allies enjoy a much stronger position than Russia, but they ought to develop new, nonlinear responses to Russian revanchism and put up a stop sign for Putin. Finally, the most pressing concern is China's military buildup, adversarial behavior, and pursuit of military and commercial applications of AI and other new technologies and standards. The key in the Pacific will be information dominance; from undersea to space, information is transforming the definition of winning.

Emerging technologies give America's competitors new capabilities and transform the character of competition, but they are no less available to

America than to others. The key issue is not so much access to these new technologies but their practical application to military capabilities. They are predominantly dual-use technologies, blurring the lines between civilian and military tech development, and are increasingly developed across borders. The key for the United States will be to leverage its vast supply of resources—human, financial, and capital—and continue its long tradition of excellence in technology development and practical innovation.

One area of focus should be the rapid development and fielding of "bleeding-edge" technologies, including AI, hypersonics, metamaterials, and directed energy. We should also improve how we incorporate and employ emerging technologies, such as some task-specific AI, cyber, and electronic warfare. Technologies alone do not mean much; innovative concepts for how best to use them do. Transforming scientific progress into real capabilities, though, requires both process and cultural adaptation.

As the discussion of China and the Indo-Pacific addressed, we should strengthen our education system and better integrate government, academia, and industry. American citizens represent a relatively small portion of the qualified and well-trained technical experts coming out of American schools, and they are the only students allowed to work in classified environments, such as the Pentagon and the defense industry. The military, and the Department of Defense writ large, can do a better job of attracting and training (and later retaining) that domestic talent. The civilian side, for example, should encourage continuing education for its technical experts and strategists, just as the military does. The military, for its part, would be wise to foster ingenuity in the ranks and allow more creative, bottom-up solutions—put the already innovative minds of the troops to use—though we recognize the attendant security risks and organizational challenges. It could also consider additional ways to allow specialized personnel—software engineers or other tech experts, for example—to rotate into the force as needed.

At the same time, better integration of the public sector with academia and industry will help those graduates make full use of their talents and renew the system of research, development, and innovation. From our perspective here at Stanford, the gap between Silicon Valley and the Pentagon looks increasingly like a chasm; the Unites States can narrow it.

Similarly, the United States enjoys a unique system of allies and partners. Our European allies lead the world in publication of papers on AI, and our allies and partners in the Indo-Pacific are numerous and capable. We would be wise to remember the value of that system and work more closely on technology development with our allies. The United States too often views military sales and cross-border data exchange through a purely business lens; they are security issues too.

A key aspect of the emerging technologies discussed herein is speed, including the speed of development, yet the government's approach to acquisitions is notoriously slow. The time it takes to develop and field a new capability is crucial and ought to be considered alongside cost and performance metrics. Congress and the administration have expanded the Pentagon's rapid-acquisition authorities for the better; flexible acquisition models will help the government reach into nondefense sectors and build better civil-military relationships with industries. Software development, for example, naturally wants to be an iterative process; often the best features and use cases are not uncovered until developers pivot in response to user feedback. More jazz than marching band, this tends to fits poorly within the Pentagon's up-front, requirements-driven model of bidding out predefined functionality as it might with complex, manufacturing-driven hardware projects. As a note of caution, though, speed is not always a good thing. In matters of national security, sober-minded strategic thinking can trump action. Rapid deployment of a new technology for its own sake will get us nowhere.

Increasing the speed of acquisitions and the scope of military capabilities requires more funding, both for the military and other agencies. High-tech research and development come with risks and require significant human and physical capital. Failure is unavoidable but a good thing. And while we pursue these emerging technologies, we recommend also addressing the very real, immediate defense challenges that confront us: a significant shortfall in military readiness, a shifting conventional balance of power, rogue states, and an assertive China.

Fortunately, new technologies may be disruptive, but they can benefit the US military in meaningful, if mundane, ways, as US Army captain and Stanford PhD candidate Katie Hedgecock explained in her panel remarks for this project to a Silicon Valley audience on campus. They can simplify and reduce the costs of logistics, the lifeblood of operations: AI-enabled predictive maintenance coupled with 3-D printing, for example, may soon reduce the long logistical supply tail of tools and spare parts needed to feed forward military operations around the world. They will likely aid battlefield decision making, easing the transition toward more dispersed, survivable command-and-control nodes. And AI may well improve personnel and talent management, reducing administrative burdens and freeing resources for operational readiness and war fighting.

The future of our relationship with China, Russia, and other actors is not foreordained; both have their own serious challenges to deal with, and it will depend on what the United States does, too. We can strengthen ourselves at home and rediscover what American leadership could do for us in our emerging new world. But the United States should also engage with those countries and work with them to build productive relationships. It's become common to point out how democracies across the world are struggling—but so are autocracies. We need better talks between the United States and China. And we need better talks between the United States and Russia. We've done this before.

Security demands a broad lens. Effective diplomacy—in cooperation with allies and partners—can secure our interests, helping to prevent military accidents, protect international commerce, and defend the option for others around the globe to pursue the same sort of democracy and self-determination that Americans already enjoy. Indeed, diplomacy and military strength are inextricably linked and necessary, complementary tools of national power. We have advocated for a whole-of-nation approach to addressing changing military technologies and capabilities. But it is important to remember that those efforts are carried out in the interests of our broader security, not necessarily military, goals. The best thing we can do for our military is to meet those strategic objectives without having to use it. Diplomacy without strength is weakness, but so too is strength without diplomacy.

For more:

- "Emerging Technologies and National Security: Russia, NATO, and the European Theater" by General Philip Breedlove and Margaret E. Kosal
- "Technology Converges; Nonstate Actors Benefit" by T. X. Hammes
- "Information: The New Pacific Coin of the Realm" by Admiral Gary Roughead, Emelia Spencer Probasco, and Ralph Semmel

*https://www.hoover.org/publications/governance-emerging-new-world/winter-series-issue-319*

**From "Information: The New Pacific Coin of the Realm"**
*by Admiral Gary Roughead, Emelia Spencer Probasco, and Ralph Semmel*

A frequent trope is that China's technological advancement is primarily the result of their stealing intellectual property. China has benefited greatly from that theft and from coercive business practices, and that execrable behavior must be stopped. However, we must accept the reality of China's prowess in technological innovation and, importantly, the application of their legitimate efforts. China's current investments and innovation in artificial intelligence (AI), microprocessors, 5G networks, quantum science, and space technology are significant, genuine, and strategic and are poised to become preferred solutions for other countries to adopt. Their innovations could end up serving as the infrastructure of the future information society and position China for technological and political advantage should its standards, systems, and policies be adopted widely by others. . . .

Our last peer competition, the Cold War with the Soviet Union, turned on nuclear and ideological strategies. The outcome of today's competition will be dependent upon information—how it is generated, obtained, transported, integrated, and used. While we acknowledge the importance of technical advancements in kinetic weapons, and China's accomplishments in that portfolio are impressive and lethal (e.g., hypersonics, undersea systems, and missile defense), technologies that define the information environment are the coin of the realm, especially in the gray zone. The information technology competition will change the nature of conflict and the definition of winning and will have profound economic, political, and military consequences. If we do not approach information in warfare as an imperative and keep our defense preparations in motion, if there is not a broad appreciation of what those preparations demand, the future way of war will be a shock to the American people.

PART

# IV

# Shared Challenges to Governance

By this point we have walked around the world, and we have considered our home in the United States. But some challenges are shared, and one hallmark of our current hinge of history is how different countries and governing systems are each reacting to the common changes coming at all of them.

Take, for example, the information and communications revolution, which allows markets and businesses to function more efficiently and individuals to learn or speak with one another at will. It has caused the social distance of space and time to collapse. What does it mean for governance? The rapid spread of information can enlighten, but it also can confuse, sow discord, and endanger democratic institutions. In the hands of autocrats, the new means of communicating become weapons of coercion, removing the information transparency that is so critical to democracy. And in the hands of individuals, these technologies enable never-before-seen forms of social and political organization, bringing new dimensions to the old problem of governing over diversity.

Let's spend some time with these shared challenges that cut across multiple arenas—challenges to media and social connection, to the act of governance in an age of hypertransparency, to the concept of citizenship in an ever more mobile world, to our health and the environment, and more—for recent events suggest that current paths are unsustainable. Hopefully, given the daunting nature of their complexity, all of these areas are in fact ones where solutions tried in in a single country can inform the progress of many, so long as we are paying attention.

To start, what can the United States and other democracies do to protect their elections from the conflict and polarization catalyzed by **social media and other network platforms**? And what are some rules of the road for confronting information warfare that might secure democratic institutions both at home and abroad? To begin addressing this challenge of information, communications, and governance, we asked two eminent scholars to contribute their thoughts.

The Hoover Institution's Niall Ferguson argued that internet network platforms such as Twitter or Facebook, left unregulated, have damaged the democratic process in America. He proposed a multidomain approach to redress what he calls an indefensible status quo.

Joseph Nye of Harvard University, meanwhile, observed that although information warfare is not new, the communications revolution has changed its very nature, making it faster and cheaper than ever before. A national strategy to address that change and secure democracy against it must both be long-term in scope and focus on resilience, deterrence, and diplomacy. The following chapters will explore both views.

It's not just elections that are at risk, though. These new technologies, alongside the other phenomena described above, are also **changing the governing process itself.** Political activists can use social media to foment opposition to policy proposals. Information has become more readily available, yet unreliable, complicating the decision making of both political leaders and citizens. Is it harder to make constructive compromises, for example, when a decision maker is judged instantaneously rather than over the portfolio of accomplishments during an appointed term? And politics becomes increasingly national as local newspapers close and politicians govern by tweet.

But these disruptive forces can also be used to make governments more effective, responsive, and honest. The connected world allows all levels of government to provision services more effectively and equitably, and it can increase transparency and accountability, for example in our primary and secondary schools. We need good governance now, and these social and political dynamics, while complicating traditional modes of governing, may help us get it.

An assortment of former and current political leaders, journalists, and intellectuals delved into this issue for our project to consider how we can improve the quality of governance in democracies across the hinge of history, starting with the United States.

Jeb Bush, the former governor of Florida and a national leader in education reform, argued that if government is to harness emerging technologies to improve its service to the American people, it will need to be fundamentally transformed. Drawing on his experience as governor, he called for greater flexibility, accountability, and transparency in government and for a recommitment to federalism. He points to education as an area of great need but also great potential for positive reform of this kind.

City and local governments are the most "consumer-facing" governance entities; when the gears at the federal level grind to a halt, local governments are increasingly tasked to take on complex social and technological questions without the benefit of a sophisticated national apparatus to support it. At the same time, ballooning pension obligations are eating up their budgetary capacity to deliver anything but basic services. Can technology help? Amanda Daflos, chief innovation officer of the city of Los Angeles, described what that mayor's office has done with these new tools, for example by improving recruiting for new police trainees.

An important buttress of good governance is the press, but Karen Tumulty, a political columnist at the *Washington Post*, attests that the spread of social media and digital platforms has transformed the fourth estate. Hallmarks of the journalism profession have gone by the wayside with digital media replacing print, news organizations consolidating, and the traditional news cycle disappearing. At one roundtable on the topic, Tumulty warned that we are still in the early stages of this transition, but that there remains a hunger for reliable reporting, both locally and nationally.

Our colleague Mo Fiorina, a political scientist at Hoover, has compared society's current moment of "democratic distemper" to that of the 1970s. Both eras, he concluded, reflect "the challenge of governing a changing population in an era of technological change." But the 1970s crisis of confidence in politics was alleviated by the decisive election of strong leaders—namely, Ronald Reagan here and Margaret Thatcher in the United Kingdom—who acted quickly but patiently to address the unique challenges of their time and who maintained enduring majorities. Will a similar solution reveal itself today?

While many of the profound changes we are now seeing are driven by our technologies, we ultimately have to react to them in human terms. The challenges of governing over diversity and adapting to technological and social changes are not new, but they are no less difficult or pressing for their ancient roots.

One key area where a crisis of confidence in governance is now playing out is in **demographics and migration**. Europe has been beset by this question, and today in the United States we continue to struggle with a framework for migration against the backdrop of our own aging society. What we have seen so far as we pass over the hinge of history, whether from the Northern Triangle of Central America or from Syria (which are relatively small regions in the scheme of potential new global migrant sources, with a combined population of just over fifty million), is the tip of the iceberg. At the same time, the United States is not full. In fact, the country needs people. So how should we proceed?

The phenomena described throughout this book—namely new means of communications, emerging technologies, demographics and the movement of peoples, and climate change—are combining to create new, daunting challenges to governance, particularly in democracies, around the world. The same regions that may be hit hardest by climate change, such as sub-Saharan Africa, will also witness rapid population growth, and problems could be compounded by poor governance. These factors together form, in the words of Larry Diamond and Jack Goldstone, a "toxic brew" and may stimulate migration for which developed countries should prepare. They argued that established democracies will likely see a potentially destabilizing surge of migration—driven by demographics, climate change, and opportunity—from less-established states. It is, therefore, in advanced nations' interest not only to strengthen their own institutions but also to help expand economic opportunity and support improved governance in those developing countries.

On the home front, James Hollifield of the Tower Center at Southern Methodist University wrote that the United States "is trapped in a 'liberal' paradox": it needs immigration to keep its economy strong, but it must also deal with the sociopolitical ramifications of that immigration. Through a review of past US immigration policies, he explained that this paradox is not new and that we can balance openness with valid concerns for security and societal values. Many regard a mature society's demographic futures, unlike economic forecasts, as unique in their certainty: "demographics is destiny." But there are exceptions. As a high-immigration nation, the United States is enviably able to shape its demography through policy decisions and governance.

As human ingenuity and drive cause new technologies to disseminate, as social interaction accelerates on the back of digital and networked communications, and as people move across the planet at historic rates, the globe itself on which all of this takes place is not static, either. One additional shift whose effects we can now increasingly anticipate with sobering fidelity is the **changing environment**. This phenomenon has introduced new health risks and challenges to an increasingly interconnected world. Extreme weather events and warming climates affect the spread of infectious diseases or even pandemics, while altering and damaging ecosystems. We further see disruptions to traditional supply chains that support modern economies. The social costs of these phenomena are rising, and individuals, organizations, and governments are struggling to adapt. At the same time, new technologies may give us new tools to address the health issues aggravated by a changing climate and even reduce some of the negative impacts of pollutants on human life.

A variety of experts at the Stanford University schools of medicine and engineering wrote on this phenomenon. One study by Dr. Milana Boukhman Trounce warned that the risk of pandemics continues to grow, driven in part by a warming climate. By reviewing past examples of responses to pandemic outbreaks, she identified effective approaches and ways technology can help these approaches work better to identify, respond to, and minimize the spread of infectious diseases. Chillingly, the emergence of COVID-19 in early 2020 proved the importance of governance and private-sector preparation and response mechanisms to that outbreak—or others yet to come. And Dr. Kari Nadeau reviewed the health effects of pollution, which have long taken their toll both here in the United States and around the world. Curbing environmental pollutants is an area where we have some positive experience in the United States, but much more remains to be done. Calling for rapid action to limit further damage to public health, she identified ways governments, clinicians, and individuals can reduce pollution exposure.

Assessing the global impact of a changing environment, developmental biologist Lucy Shapiro and physicist Harley McAdams meanwhile described how human-driven increases in carbon dioxide have created a major biological disruption. They traced the effects of that disruption to some unexpected quarters, from changes in the world's oceans to the spread of animal and plant pathogens to new areas. And bioengineer Stephen Quake offered examples of the groundbreaking work being done here at Stanford to address these very health and economic effects, identifying scientific and technological advances that may help humankind adapt to a changing environment.

Building resiliency in the face of a changing environment poses novel governance challenges. Fortunately, much as with demographics, climate science actually allows us to look forward to basic but important long-term changes in our environment and weather with surprising acuity. "Resiliency" scholar Alice Hill examined whether governments have mechanisms to take advantage of this information. She argued that US federal, state, and local bodies should develop new capabilities and planning processes to reassess how and where public infrastructure is built, given changing environmental risks, and offered a series of concrete policy and governance reforms to promote resiliency.

We close with perhaps the most daunting challenge, but in many ways an almost forgotten one: that of **nuclear weapons**. Nuclear weapons pose a unique, existential threat to humankind. For over half a century, states and international governing bodies have carefully managed that threat. We have had real success in limiting the proliferation of weapons and material and restricting arms programs, but sustaining, much less expanding, those successes is becoming increasingly challenging due to twenty-first-century technologies and global instability.

Nuclear weapons pose great dangers, but as former secretary of energy and CEO and cochair of the Nuclear Threat Initiative the honorable Ernest J. Moniz wrote for this project, nuclear power can be a great boon for humanity as the world comes to terms with the very real problem of climate change. He wrestled with "the inherent dual-use nature of the nuclear fuel cycle" and argued that elimination of nuclear weapons combined with a major global expansion of the civil nuclear fuel cycle will be particularly challenging. Squaring this circle, what he calls "the fundamental test of the nuclear era—balancing nonproliferation with peaceful uses of nuclear energy" will require new methods of detection, prevention, and rollback as well as new means of addressing subnational risks. Fortunately, Secretary Moniz sees promise in the emerging technologies and policies available to those working on this problem and suggests a new path for dealing with the North Korea challenge. If the United States chooses to step up quickly and lead nonproliferation efforts, it can better manage the risks of nuclear weapons while realizing the potential of nuclear power.

We also looked at what may be the most dangerous nuclear hot spot in the world: the Indian subcontinent. Ashley J. Tellis, a senior fellow at the Carnegie Endowment for International Peace and former senior adviser to the US ambassador to India, contrasted the Pakistani and Indian nuclear programs but warned that neither state is satisfied that they have sufficient weapons inventories to ensure their national interests. Both parties will, he expects, continue to increase their stockpiles, but the relationship does not have to spiral out of control. To secure a more stable future, Tellis called on the international community to counter Pakistan's coercive behavior and help "break the linkage between political revisionism and nuclear weaponry in ways that will ultimately assist both Pakistan's internal stability and the orderly evolution of India and Pakistan as responsible nuclear powers."

In addition to these two excellent contributions, we were joined on our discussion of the changing nuclear threat by our friend and colleague former senator Sam Nunn, cochair of the Nuclear Threat Initiative with Secretary Moniz and a visiting fellow at the Hoover Institution. Senator Nunn discussed the unique nuclear dangers presented by emerging technologies and geopolitical uncertainty, and what the United States should do as it embarks upon its own weapons-modernization program.

Let's go deeper into what sort of options we may now be able to make out on the horizon for each of these shared challenges to governance.

CHAPTER 9

# The Information Challenge to Democratic Elections

Prior to the introduction of the internet, posited historian and political scientist Niall Ferguson in his contributions to our project, "only the elite could network globally." Now nearly two and a half billion people do so on social media platforms. At the advent of this networked age, most practitioners pictured a better educated, more knowledgeable populace enlightened by the democratization of information. The public would have up-to-the-minute access to information and the ability to communicate globally. Voters would have more ways to learn about candidates and engage in political speech than ever before.

That Panglossian view of the internet proved simplistic. The spread of information and new means of communication—particularly social media and other network platforms—created new vulnerabilities in democratic states. Harvard historian Joseph Nye explained that while authoritarian regimes can manipulate or even control information flows, democracies, in their commitment to transparency, find themselves on the defensive in this realm.

Foreign actors can manipulate information, particularly in the cyber domain, to undermine trust in institutions, sow domestic discord, encourage partisanship, or otherwise complicate electoral processes, as the Russians demonstrated in their interference with the 2016 US presidential election. But such behavior is not the sole purview of foreign entities. Private citizens and corporations alike have the power to influence election outcomes in new and powerful ways.

If the 2016 election showed the American public the potential of network platforms as tools of political manipulation, it also taught us how thorny the problem is. Although we do not know if Russia's interference actually affected voters' choices, it achieved an arguably more important goal: it undermined faith in the American electoral and political process and in this country's democratic reputation.

Our contributors to this project tackled two separate but related issues: 1) the domestic problem of managing highly powerful network platforms, and 2) the international problem of information warfare enabled by these new communications technologies. The former demands a reconsideration of US policy at home: the current status quo of "self-nonregulation" by the network platforms—alongside a lack of responsible behavior by some of their users—has proved wanting. The latter requires both US policy corrections and multinational engagement: as Harvard historian Joseph Nye argued, we can improve our resilience to foreign information campaigns and our capacity to deter them while also engaging in diplomacy to define new rules of the road.

## Cyber-Information Warfare

Information warfare is an old form of competition, and one practiced by friends and foes alike. The British cut Germany's overseas communications cables at the outset of World War I, but they also fed the United States the Zimmermann Telegram to encourage US engagement later in the war. As old as it may be, though, new technologies have made information operations faster, more effective, and cheaper.

Russia's interference in the 2016 election comes to mind again: 126 million Americans saw posts generated by the St. Petersburg–based "Internet Research Agency." In the 1980s, the Soviet Union's Operation *Infektion* conspiracy fed KGB disinformation opinion pieces about AIDS to newspapers around the world claiming that the disease was an American bioweapon—it took four years to spread into mainstream media. In contrast, the 2016 Comet pizzeria conspiracy theory, in which an anonymous Reddit poster detailed a fictitious sex-trafficking ring supposedly based at a Washington, DC, pizzeria—and which led to one North Carolina man arriving with rifle in an attempt to break it up—spread across the country in a matter of hours, with posts amplified by both domestic and foreign social media accounts. While the national origin of this modern conspiracy theory is undetermined, it illustrates how the cost of creating a Facebook post or effectively generating other online content is microscopic compared to that of traditional human intelligence operations.

In his own work, Joseph Nye drew the distinction between soft and sharp power. Soft power, he argued, rests on persuasion, while sharp power involves deception or coercion. Soft power is exercised openly, sharp power covertly.

The openness of the American system makes it more vulnerable to sharp power than more closed systems are, and states and nonstate actors alike have a host of tools available for disrupting democratic processes: manipulation of voters through fake news, targeting of candidates anonymously or under false names, creation of inauthentic groups to generate conflict, and sowing of chaos and disruption. Russia, China, North Korea, Iran, and others all wield these tools against the United States.

However, while Russia proved adept at exercising sharp power, Russia and other authoritarian states, including China, are less adept at soft power. Russia's actions in the 2016 election, for example, fall under the umbrella of sharp power. The Russian news channel RT, on the other hand, generally engages in the aboveboard exercise of soft power. An American soft power analog would be Radio Free Europe and Voice of America, which were powerful tools of information warfare during the Cold War.

New technologies—chiefly social media and other network platforms—are fertile ground for the exercise of a cyber-information war. They can promote polarization and spread fake stories. The business of Facebook, YouTube, or Twitter is to maximize the attention of their users. The more time and attention given, the more advertisements seen, the more ad money for the platforms. False stories, outrage, and emotion capture attention far better than sober-minded articles or videos. It is easy to believe, as some have argued, that YouTube's algorithms tend to suggest videos that push viewers toward more extreme ends of the political spectrum.

As modern technology, including artificial intelligence, make the manipulation of images and videos easier, bad actors can create fake or altered content that is increasingly difficult to distinguish from its authentic counterparts. Introduced late in a campaign, such altered images could spread quickly enough—before Facebook's operators, say, could take them down—to influence the outcome of that election.

Russia has learned to weaponize social media and use it as a tool against the United States. The Internet Research Agency and similar operators have created fake accounts, catalyzing polarization in American society, and amplifying extreme voices on both sides of the aisle in the United States.

## How to Protect Our Democracy from Foreign Interference?

Social media and network platforms have come to dominate the public square, enabling broader and more complex social networks and political organizations than ever before. Information has always been an extremely valuable asset—once costly to obtain and share, now essentially free to all strata of society. But the spread of internet platforms comes at a cost. Whereas social media was once seen as a tool to disrupt non-tech-savvy authoritarians, we are increasingly aware of how it can be manipulated to transform democratic elections too.

Democracy depends on transparency and the spread of open, trustworthy information. That commitment to openness is a vulnerability, but it is also a great virtue and a great value, one deserving of protection. When considering what can be done to redress the information challenge to governance, then, the United States must commit first and foremost to doing no harm to its democracy. We must be careful not to sacrifice it.

In other words, the US government should not try to stop transparent information campaigns—legitimate exercises of soft power—in its effort to secure itself against illicit interference. Nor should it look to technology companies to solve the challenges. Facebook, after not seeing the problem coming in 2016, has taken steps to address it, hiring new employees and applying artificial intelligence to find and remove hate speech, bots, and false accounts. But the enormous quantity of content, the entanglement of foreign- and domestic-generated content, and the mix of human and bot actors complicate the problem; the vast majority of Russian posts during the 2016 election amplified existing content created by Americans. Moreover, just as AI can help monitor and police content, it can also be used to generate new, harder-to-identify false content. The technical contest between the network platforms and foreign agents is likely to remain a cat-and-mouse game.

In our discussions, Nye proposed a threefold approach, which we believe is wise: The United States should look to increase resilience at home, strengthen deterrence, and engage in diplomacy with foreign powers.

*Resilience:* The United States must take steps to harden its electoral and political systems against cyber-information warfare. American governmental and nongovernmental institutions, including academics, could upgrade the security of US election infrastructure by training local election officials and improving basic cyber hygiene, such as using two-factor authentication. Given how much political campaigns and electoral offices rely on interns, volunteers, and other part-time workers, it may be difficult to train everyone, but even some training and better resilience would make a difference.

The United States should also encourage the development of higher standards in software by revising liability laws and encouraging development of the cyber insurance industry as the number of points of vulnerability to cyber intrusion expands exponentially with the internet of things. Election laws could also change to force candidates to put their names on online political ads just as they do for television ads, and to ban the use of bots by political parties or campaigns. As in other areas of cybersecurity, improved information sharing between government and industry would contribute as well.

*Deterrence:* Deterrence can further be established through the threat of punishment and entanglement. Effective punishment would, of course, depend on reliable attribution. Deterrence need not be perfect to be useful, but it ought to raise the cost for malevolent actors. Cyber intrusion could be treated as we do crime. When seeking to deter criminal activity, the certainty of getting caught matters more than the severity of the punishment, so better, faster attribution and action will be key. The Trump administration's September 2018 executive order promising sanctions in response to election interference is a step in the right direction. Entanglement complements punishment; if an attack on the United States hurts the attacker, that reduces the incentive for malicious behavior. Together these approaches could help develop shared, international normative taboos around the use of information-based foreign interference. Deterrence won't solve the problem, but it could increase the cost and difficulty—defensive measures could then be better focused on those attacks that do still occur.

*Diplomacy:* This arena is not conducive to conventional, binding international arms-control frameworks, such as those employed in the Cold War to limit nuclear weapons technologies. For example, a Twitter account is inherently dual use, because it is both a tool of disinformation and a means of innocuous social networking; the key variable is the user and the user's intentions. Instead, we ought to establish rules of the road to limit certain malicious behavior.

We are proposing not a treaty but a set of agreements or understandings, which will depend on the values of the involved parties. Just as the United States and the Soviet Union came to the 1972 Incidents at Sea agreement to reduce the risk of inadvertent crises, so too could the United States and Russia conceivably commit not to interfere covertly in elections while allowing overt broadcasting and transparent information. Each side could unilaterally propose and share its own expectations of conduct, tracking and communicating how the cyber behaviors it observes over time do or do not comply with those expectations. The United States does not have to act alone here. It could work with its allies and partners—fellow liberal democracies—to coordinate collective action: sharing defensive recommendations, mutually shoring up electoral and political processes, and collaborating on diplomatic agreements.

In other words, the United States can work to establish upper bounds of cyber-information activities—thereby allowing US officials and others to focus their resilience-building efforts on a narrower range of challenges—and prepare for prompt retaliation for activities that exceed these bounds.

## What to Do about Network Platforms?

As the United States addresses cyber-information warfare, it ought to also consider the preeminent power of its own domestic network platforms. Manipulation of information to disrupt our electoral process demands a response, but the information challenge to governance extends beyond cyber-information war.

For better or worse, internet companies now hold immense political power. As social media analyst Robert Epstein has documented, they can shift election outcomes and public opinion through slight manipulation of search results or suggestions, content feeds, and other user interfaces. So-called "dark patterns" are a well-documented aspect of digital interaction design outside of the political arena and are an emergent threat here too: almost imperceptible tweaks to underlying algorithms can swing voters toward a single candidate.

Niall Ferguson ably describes the current status quo: eight technology companies—including Facebook, Alphabet, and Tencent—dominate global internet commerce and advertising. They are immensely profitable. To Ferguson, network platforms have become a "public good," not just commercial enterprises, trading on the attention of the public. But they are contaminated with fake news and extreme views, some incited by our nation's adversaries. And network platforms, such as Twitter, have transformed governance in the United States.

They may be public goods, but these platforms are essentially self-regulated, or more accurately, self-nonregulated. What regulation exists gives the network platforms significant leeway. Under US federal law, they are generally not regarded as publishers, nor are they liable for the content they host, or the content they remove.

It is unsurprising, then, that companies curate and customize content on their platforms. As described above, they seek to maximize user attention and have done so to great effect—the average American spends 5.5 hours per day on digital media. Into this environment come fake news and polarizing content, which spread more quickly and attract more attention than sober-minded alternatives.

With their vast network of users and grasp of user attention, US internet platforms became a key battleground of the 2016 election—one in which the winning campaign was most focused.

## Regulation, Firewalls, and Other Proposals

Should anything be done to change the imperfect power held by internet platforms? Two foreign models for managing internet platforms suggest what not to do.

Europe has adopted a tax, regulate, and fine model. To Ferguson, Europe "seeks, at one and the same time, to live off the network platforms, by taxing and fining them, and to delegate the power of public censorship to them." China, on the other hand, zigged when the West zagged, adopting "internet sovereignty" in contrast to the US-led internet freedom agenda. It built the great firewall and fostered its own domestic industry through total protectionism.

Neither foreign model appeals. In the wake of the 2016 election, US network platforms responded—within their own largely laissez-faire business environment—by pledging more-strenuous self-regulation at the firm level. Facebook, for example, now requires disclosure of who pays for political ads, uses artificial intelligence to detect bad content and bad actors, removes certain foreign government accounts, and reduces access to user data. These are measurably helpful but remain essentially reactionary steps. It is hard to have confidence that they have solved the next threat.

So while we do not know the precise solution to these problems, let us consider a few other options. It is easy to focus on the new problems generated through these platforms while taking for granted the informational value and personal satisfaction they also generate. We therefore wish to redress the more damaging effects of these technologies while continuing to take advantage of their promise.

Ferguson has recommended that the US government make network platforms liable for content on their products—essentially scrapping the 1996 Telecommunications Act provisions protecting them—while also imposing First Amendment obligations on them. That is, the platforms themselves should not be allowed to decide what speech is acceptable by their own rules. His approach would give users and competitors recourse to challenge companies in the courts.

But as our discussants noted, there are certain internal contradictions to such a proposal. Asking companies to monitor content for which they could be held liable—a task that would necessarily rely on both AI and large human teams—would likely complicate their ability to post everything permitted by the First Amendment. And what of anonymity, which has been so crucial to the internet freedom agenda? How do we protect anonymity while also enforcing liability? Perhaps a first step would entail banning content generated by bots or nonhumans.

Alternative, but unappealing, regulatory steps would include using the Federal Communications Commission's 2017 scrapping of net-neutrality rules to encourage internet service providers themselves to monitor the digital content flowing through their pipes, but there is little reason to believe they would do a better job given their own profit incentives. Antitrust efforts intended to break up platform companies would also likely be of limited utility: these efforts would be slow and of questionable effect, and run against the natural winner-takes-all direction of network platforms. Moreover, we must remember that historically regulation tends to cement the dominance of the largest players, stifling innovation and competition. Witness, for example, the growing market share of large integrated national banks over community and start-up banks following the onerous compliance regime of 2010 Dodd-Frank financial regulations—only a handful of new bank charters were issued from 2011 to 2018, compared to hundreds per year before the regulations, and during which time the number of small US banks fell by one-third.[26]

The government and public could ask more of the tech industry, through new authorities that could backstop self-regulation. Government or civil organizations could press companies to be more transparent about their criteria for managing content on their platforms, while also establishing a recourse to challenge network platforms' actions. Along those same lines, companies could be ordered to make some portion of their algorithms available for public review or for the review of a select court, in the vein of the Foreign Intelligence Surveillance Act (FISA) court.

The public itself—users—have an important role to play as well. As both creators and consumers of the content that populates network platforms, individuals can refrain from relying on social media for "news" and be discerning in what they share. What are their responsibilities to each other? Appeals court judge and Hoover fellow Michael McConnell has argued that liberal democracies like the United States, with a diverse and expressive citizenry, must be voluntarily committed to a "much thicker understanding of the common good" than subjects of unitary authoritarian states. At that same time, as free peoples lacking overarching institutions of state power—control of the press, a ruling lineage, a unifying race or religion, and so forth—it has been "both more necessary, and more difficult" for these democracies to establish a shared idea of citizenship. Network technologies only intensify this original "flaw" of democracies, one that the United States has successfully managed for more than two hundred years of change through evolving concepts of public and mutual responsibilities. If this is indeed the new town square, then what does digital citizenship look like in an emerging new world? What are the responsibilities and privileges that it entails, can they be lost, and from where do they arise? How are they taught? The government and the companies do not bear sole responsibility for addressing this challenge.

In the United States, some may look to the past, when symbols of public trust—Walter Cronkite and Huntley and Brinkley being the canonical examples—gave us the news. Those days have passed, but the importance of trust and reputation remain. Internet companies would be wise to regain the trust of the people through careful stewardship of their platforms, giving priority to accuracy over attention, and willing public-private engagement. And users, with their immense agency, must redefine what it means to be self-responsible citizens in this realm as much as they do in the other 18.5 hours of their day.

For More:

- "What Is to Be Done? Safeguarding Democratic Governance in the Age of Network Platforms" by Niall Ferguson
- "Protecting Democracy in an Era of Cyber Information War" by Joseph S. Nye

*https://www.hoover.org/publications/governance-emerging-new-world/fall-series-issue-318*

### From "What Is to Be Done? Safeguarding Democratic Governance in the Age of Network Platforms"
*by Niall Ferguson*

Once upon a time, only the elite could network globally. David Rockefeller—the grandson of the oil tycoon John D. Rockefeller—was a pioneer networker. According to a recent report, "He recorded contact information along with every meeting he had with about 100,000 people worldwide on white three-by-five-inch index cards. He amassed about 200,000 of the cards, which filled a custom-built Rolodex machine, a five-foot-high electronic device." Rockefeller's contacts ranged from President John F. Kennedy to the shah of Iran, Pope John Paul II, and the astronaut Neil Armstrong. Henry Kissinger generated the most cards—thirty-five in all, describing hundreds of encounters over sixty years. Not far behind was Gianni Agnelli, the Italian industrialist who ran Fiat. In his memoirs, Rockefeller recalled how, while serving as an Army intelligence officer during World War II, his "effectiveness [had] depended on my ability to develop a network of people with reliable information." It was an experience that he took home with him when he joined Chase Manhattan after the war.

What was once the preserve of a tiny elite is now available to everybody with an internet connection and a smartphone, tablet, or laptop computer. Facebook was founded at Harvard in 2004, before David Rockefeller's ninetieth birthday, and rose rapidly to become the world's dominant social media platform. The company's foundational premise was and remains that "simply through sharing and connecting, the world gets smaller and better." Connecting people on Facebook, Mark Zuckerberg declared in 2015, was building a "common global community" with a "shared understanding." . . .

Speaking at MIT in February 2017, Obama suggested that "the large platforms—Google and Facebook being the most obvious, but Twitter and others as well that are part of that ecosystem—have to have a conversation about their business model that recognizes they are a public good as well as a commercial enterprise." It was, he said, "very difficult to figure out how democracy works over the long term" when "essentially we now have entirely different realities that are being created, with not just different opinions but now different facts—different sources, different people who are considered authoritative."

Obama was right that we have a problem, and it is not a problem we were prepared for by the masters of Silicon Valley. "I thought once everybody could speak freely and exchange information and ideas, the world is automatically going to be a better place," Evan Williams, one of the founders of Twitter, told the *New York Times* in 2017. "I was wrong about that." Indeed, he was.

CHAPTER 10

# Governing over Diversity in a Time of Technological Change

The information and communications revolution has complicated governance everywhere. It has broken down traditional borders: people can communicate, organize, and act both with their fellow citizens and across country boundaries. The age-old challenge of governing over diversity grows more difficult by the day.

We've considered the potential impacts of these technologies on democratic elections. After any election, though, one must govern. And good governance is in short supply. Look around the world: Washington, DC, is paralyzed by partisan fights and unable to fund the government on time; France is ruled by protest, as seen in the Macron administration ceding to the demands of the yellow vests; Chile, one of Latin America's few enduring bright spots, is seized by constitutional crisis, process and first principles consumed by negotiation with the mob; in Germany, Angela Merkel, once a model of European leadership, has lost her support; and Brexit consumes the United Kingdom and its parliament—once steady, even the British are coming apart. And the problems go beyond just Western states. Supranational institutions—the European Union for one—are struggling, and autocratic governments in Russia, China, Venezuela, and the like have either run their countries into the ground or are dealing with economic and demographic headwinds of their own making. Are these new information technologies to blame?

## Twenty-First-Century Technology Complicates Governing

Two factors have contributed to governments' sclerosis: the predictable growth of the administrative state and the unpredictable disruption brought by technology.

Social media, network platforms, and the modern internet have changed the relationship between government and the public. They have, as former Florida governor Jeb Bush argued in our discussion on this topic, "radically transformed our private and work lives by making information more accessible, communication faster, and businesses more competitive." By enabling the nearly instantaneous spread of information and the mass coordination of disparate peoples, these technologies, in particular social media and online news, have upended both government communications and the traditional journalism industry. And, rather than encouraging discourse and a diversity of opinion, they feed discord, empower protest, and nationalize our politics.

Communications is a central aspect of governance. It used to be that, in government, if you did something good or you solved a problem and the public noticed it, then you would get political support. The first step was to solve the problem, but you then needed to make sure constituents understood what you had done; you needed an organized communications system to get the message out. It was a positive feedback loop, and it encouraged those in power to seek effective, positive results. The advent of new communications technologies has broken that feedback loop. Social media is unedited and off the cuff. It lends itself more to complaining or protesting than to communicating in-progress or successful efforts. To be sure, government officials can use these platforms for positive messaging, for recruiting, or for other such purposes, but even then, any mistake can go viral and undermine the effort.

More broadly, social media feeds criticism and discord in politics, not the exchange of ideas. We have discussed this effect within the context of elections, but it is just as real after the election. One look at Twitter during the 2019 impeachment hearings will show that social media platforms are not effective venues for the pursuit of compromise or consensus. Yet that pursuit is what governing is all about.

Social media users often are in the business of expressing quick, simplistic opinions, mostly criticisms of other people's ideas. That makes them particularly effective at shutting down policy initiatives. It is easier to organize large affinity groups on Facebook in opposition to policies than to develop those policies and put in the hard work of implementing them. In particular, "one-to-many" structured social media allows for a small number of particularly exercised parties, at low cost, to at least appear to enlist a broad collection of aggrieved supporters. We say they appear to do so, because the casual act of a retweet or a like or a share is not the same thing as a durable personal belief—or a vote. As *Wall Street Journal* columnist Dan Henninger wrote in a June 2019 column, "the normal give-and-take of governance isn't working anymore because factions can leverage social media to strangle any proposal in its crib."[27] Those factions are not the localized ones imagined by the founders but national collections of people.

Indeed, the instantaneous spread of information has had the odd effect of nationalizing American politics. Media and institutional forms have become linked. For example, the advent of American broadcast television in the 1950s ushered in the creation of national, consumer-focused megabrands who could afford to advertise a standardized product to undifferentiated masses of viewers. Television needed advertisements, and advertisers in turn fit their own products to television. But, as others have noted, when the public square went online, politics too became national in scale, yet personal in scope. As opposed to television's appeal to commonality, today's internet offers the ability to generate thousands of thin-but-broad categories of informational flow along topical lines. And for better or worse, the miracle of communications technology improvements means that separation through space creates no cost to participation: retail politics is nationalized. Trending Twitter hashtags are countrywide themes, not a conversation among neighbors. What happens in small-town Arizona, for example, can become news in New York City almost immediately, and the collapse of local journalism is both a product and driver of this effect.

Karen Tumulty, a columnist for the *Washington Post,* enumerated for us the decline in local news organizations and the attendant loss of accountability. Due primarily to social media and digital news platforms, circulation numbers have shrunk to their lowest level since 1940. The loss of local papers can be seen in congressional press galleries. These press galleries were once populated by correspondents from all over the country, who reported on what their local representative was up to in Washington. Their reporting enforced a certain accountability in congressional offices; but now, constituents do not have the same awareness of what their elected representatives are, in fact, doing. The current news system rewards international, political, and sports news. National politics, in other words, has trumped local politics in the journalism business.

What does it mean for politics to go national like this? In the words of one of our roundtable participants, political writer Christopher DeMuth, "We have entered a world of empowered mass intimacy" and can now "organize ourselves into highly defined networks of affinity and endeavor."[28] Those networks can bring many benefits—consider the revolutionary idea, which Airbnb has made common, of letting strangers stay in our homes—but they also, DeMuth warns, are "fracturing our politics." They feed political polarization by giving a national airing to extreme grievances, which creates distorted images of those on the other side of the aisle and in turn feeds distrust. Moreover, while social media and other communications technologies have lowered the costs of political mobilization and activism, Americans have increasingly focused on discrete political concerns. They look to Washington to solve their problems, so we get small national factions capable of gumming up the workings of government and stopping policies they do not like.

In all this, the basic institutions of American government have suffered. Christopher DeMuth noted that the establishment of Congress was a political innovation. It allowed representatives from all over the country to gather together, understand the problems in other parts of the country, and discuss national concerns. But, as already discussed, we no longer need congressional sessions to have a national conversation, nor do we need representatives to share the concerns of western Pennsylvania with eastern Arkansas. In other words, he argues, Congress has lost its primary reason for being.

Government has become less representative and more administrative, as DeMuth described in the *Claremont Review of Books*.[29] Congress relinquished its authority to executive agencies with time, and it redistributed power within its own halls. Committee chairs have lost much of their power, and both chambers of Congress have been "democratized." Legislative direction lies in the hands of leadership, while members act as individual political entities, always looking out for their own electoral survivability, and therefore focused on their own personal celebrity. The legislative body does not do much legislating. It passes occasional sweeping legislation along party lines, such as the Affordable Care Act or the Tax Cuts and Jobs Act, but struggles with more mundane exercises, such as passing on-time funding bills and performing effective oversight. No longer a home for what DeMuth described as "muddled decision making," Congress has given law- and rule-making power to what he terms "declarative" administrative agencies in the executive branch. Unelected bureaucrats now make rules governing all manner of daily aspects of life, and the public has little choice but to accept those rules, at least in the near term. All told, the balance of power in Washington has become unbalanced: legislators have traded deliberation and legislating for individual pursuits, fundraising, and lobbying the executive office.

It may be unsurprising, then, that we see the personalization of politics and the weakening of political parties. Our colleague and founder of the Israel Democracy Institute Arye Carmon has described how the costs of personalization are not limited to the particularities of the US democratic system. A revolution has also occurred in Israeli politics, with individual campaigners or policy makers seeking to appeal to niche motivations. When the public begins to vote for specific individuals, with little concern for party platforms or principles, it leaves the parties open to demagogic capture and damages accountability. Government ends up being whipsawed between personalities, without consistent policies or institutions to guide policy.

The distended and paralyzed government we have today is, to borrow a phrase, the high cost of good intentions. After the destruction and violence of World War II, nations, in particular the United States, were presented with the very real task of preventing anything like that from happening again. To do so, they established strong national and international governing institutions—Bretton Woods for global finance and trade, NATO, powerful defense capacities—and the United States played a guiding role in all of it. Built to be strong and durable, it should come as no surprise that both national governments and supranational governing organizations became more powerful. In the United States, we can see how the administrative state became more and more influential. With time, it established new programs, accrued new responsibilities, and solved new problems, but now it may have reached its limit.

Government seems incapable of addressing immigration, health care, or infrastructure, and politics is increasingly disconnected from the reality of governing—it's normal to critique the outrageous campaign promises that politicians of all stripes make. Today, though, the same attitude seems to permeate governance even once the election is over, no matter how unfeasible. As alluded to above, the incentive structure for individual politicians, driven in large part by technology, feeds this disconnect; if the goal is social media engagement and news coverage, then it makes sense to espouse "big ideas" at the expense of process and reality. The press release about legislation introduced has shoved aside the pursuit of actual legislation passed. The positive feedback loop of governance—wherein governments developed good policies, implemented them, solved problems, and were rewarded for it—has been broken. In its place we hear about all the things governments can and will do, but no problems get solved. We now have a negative feedback loop of public anger, more outlandish promises, disappointment, and then more anger. In the words of Dan Henninger, "We can now stop anything we don't want, but can't enable anything we need."

## Restoring Good Governance in the United States

This new normal is unacceptable. Governments do not govern, and there is a serious risk to a lack of governance. It leads to bad policy, which, without the anchor of durable institutions, can in turn lead to things rapidly falling apart, as we have seen across Latin America. Civilization has a thin veneer. Moreover, the new normal spans national boundaries, leaving a governance vacuum across the democratic world, even as rapid technological and demographic changes transform societies around the globe. To repeat, the world needs good governance, and the United States is the one country with the capacity and principles to help it get there. So the onus is on the United States to get back into the business of governing itself.

There is no one perfect solution or grand bargain to fixing American politics and democratic governance. China is attempting to establish an alternative model for how to govern over a large, diverse population. But, as described earlier, the People's Republic of China has serious problems of its own; it is not an attractive model, nor should we aspire to be more like an authoritarian state. As just one example, the escalation through the summer of 2019 of a relatively minor governance matter—a criminal extradition agreement—into a near-existential breakdown of debilitating and violent daily protests in Hong Kong was the result of repeated interference by mainland Chinese government minders who refused to let local leaders compromise with Hong Kong residents. The experience showed how well the CCP's governance model handles social disagreements without the easy threat of compulsion hovering in the background: poorly.

For the United States then, the first step, instead, is to get smaller. We seek a new federalism. Political discourse focuses excessively on national grievances too large to be solved all at once, so let's lower the stakes. Return power to the states, districts, and local municipalities, and seek out ways to make governments more efficient, responsive, and accountable through the integration of new technologies.

Governor Bush warned that we have a twentieth-century governance model in a twenty-first-century world. The way government is organized at every level has not changed, but society and the role of government have. The bureaucratic, centralized government described above is not prepared to realize the opportunities presented by new technologies. To Bush: "Our complacent monopoly of a government needs to be brought into the twenty-first century. We need to cut regulations, make civil servants accountable, fix broken public procurement processes, and, for the sake of accountability, experimentation, and local autonomy, return power to state and local governments."

Governor Bush's accomplishments in Florida are a testament to the power of thinking smaller and holding government accountable. He inherited an overly large and protected civil workforce, a broken budgeting system, and a struggling educational system. And then he reduced government payrolls while increasing the quantity and quality of services provided. Florida transitioned toward a defined-contribution benefits plan for civil servants, and Bush championed a pilot program to overhaul the state's Medicaid program in an effort to reduce runaway costs. Most notably, he helped lift Florida's K–12 education out of the doldrums, turning it into one of the country's better school systems.

The principles Governor Bush applied to reforming education are instructive for other areas of government. His administration leveraged the state's control over the school system to increase transparency about student and school performances, set clear baseline expectations, and hold schools accountable, in part by determining funding against the performance baseline. Doing so required a willingness to disregard broken processes and shatter the status quo, but also the political courage to accept the risk of pushing for reforms. It also was made possible through the collection and careful use of data as a means to increase transparency and accountability, which created political support for his policies.

Government services, for the most part, are delivered at the state and local level. Citizens, in turn, interact most closely with those more local echelons of government. It is at that nexus of the governed and the government where we see real opportunity for better governance.

The United States has an opportunity to improve the functioning of government in mundane but significant ways. Its leaders can do so by applying the principles Governor Bush presented and by employing new technologies to understand their constituents better and to improve the quality of services they provide. The widespread availability of data on citizens gives governments access to more and better information than ever before. Amanda Daflos, the chief innovation officer for the city of Los Angeles, described how her office has taken advantage of that information. Working with individual departments, they have introduced better online interfaces for government service and enforcement agencies, for example by making it easier to pay a ticket or renew a driver's license and by designing automated chat bots to give residents information about available services. She and her colleagues have also sought to improve hiring processes and attract talent to government jobs—like Governor Bush, she emphasized the importance of getting good people into government.

Of course, improving governance through technology is easier said than done, but we have learned some best practices. Amanda Daflos's team always develops each new solution or method in cooperation with the department that will have to implement the new process. After the first year of development and work, the innovation office then turns over responsibility for the new initiative to that department. In other words, they get the buy-in and feedback of the operators early, echoing a motto of the Reagan administration: "If you want me on the landing, include me on the takeoff." Moreover, by making these evolutionary innovations at the city level, her team is able to understand their constituency well enough to respond effectively; they recognize that the public is the customer, and like any good service provider they seek to respond to that customer. And, though hardly small, the Los Angeles mayor's office is a more agile institution than the federal government.

It is easy to identify problems for governments to fix—and we certainly did a good deal of that already. But to solve a problem, government first has to understand the components of that problem, recognize which agencies or entities have the authority to address it, develop solutions, and then establish programs to implement those solutions accordingly. And as "mayor for life" and longtime speaker of the California State Assembly Willie Brown reminded us in his own reflections as one of this country's most savvy political dealmakers, it means creatively cutting deals—however unsavory that may sound to the technocratically inclined. Putting together a successful political constituency is ultimately about showing policy makers how their own successes are tied to the delivery of successful governance. It is a complex process. Consider a homelessness crisis in a city like San Francisco. If the mayor's office wishes to do something about it, it will need to understand the different types of homelessness in the city and all their causes. It will need to establish durable and affordable programs that deal with those different challenges. It will need to get the public on its side by showing them what success looks like. And with that support in hand, it can work across all levels of government, including state authorities, to implement.

Technology can help with some steps in the process, particularly in understanding the problem. But even Mayor Brown—who presided over "dot-com" San Francisco as recently as 2004—conceded that his successful style of political machination would not be possible in today's world of social media. Some of the solution comes down to having healthy institutions with healthy policy-making procedures. We described the devolution of the US Congress above, citing Christopher DeMuth's work. It no longer possesses an effective legislative process in part because its institutional structure has been dissolved. And now we have runaway federal deficits and debt and inconsistent funding for vital national priorities, especially the national defense. The architects of those institutions should consider what reforms they might explore to encourage a new generation of leaders in this changed technological environment whose own political success can once again be aligned with the delivery of good governance. The public is waiting for an answer.

In the meantime, to the extent possible, we should take our futures back into our own hands and commit to restoring the principles of subsidiarity on which this country was founded. Recognizing that the federal government reserves responsibility for certain functions—again, providing for the national defense comes to mind—we call for states and local municipalities to think smaller and to seek, as much as possible, to solve their problems as close to home as possible.

## Bias

A word on social bias in emerging technologies, the subject of substantial roundtable discussion and disagreement. Discussants agreed that machine learning and other algorithmic decision-making systems of the sort now regularly used by both internet companies and other institutions, including the government, are biased, reflecting bias in the data used to train the systems, in the design of algorithms, and in the interpretation of results. Internet companies, for example, collect personal information, create and refine behavioral profiles of individuals, and develop algorithms that curate content, all with the objective of maximizing attention and commercial profits. So it is often in the firms' economic interest to discriminate in terms of users they market to, or in the content that they display to them. And discussants readily accepted that such decision-making systems can perpetuate or even introduce new bias. Some forms of bias are illegal, such as when ad buyers target housing options in certain locations to one race and in other locations to another race. Legal discrimination might include showing quality investment opportunities to one class of users and shady schemes to another. And bias can also be an issue in many other contexts as well outside of internet advertising or content curation, such as machine-learning systems that support decisions on hiring, mortgages, insurance availability and rates, and sentencing guidelines.

The disagreement among the discussants was about the scale and importance of such bias within an overall system, what might be done to effectively counter it, and the collateral costs of doing so. While some argued that any amount of unintentional or illegal bias is problematic, and saw potential to generate machine-learning decision-making systems without the biases common to human decision making, others pointed to the known biases of humans in our own decision making, which suggested comparing machine-learning systems to a status quo (flawed) human baseline rather than an idealistic one of perfection.

And whereas some participants offered (federal) government agency regulation or congressional lawmaking as ways to provide enforcement strong enough to encourage powerful technology companies to actually comply in removing bias, others warned that involving government agencies, who have less information and technical expertise available to them than private firms, could actually end up making bias problems worse in a fast-changing technological environment. These experts instead offered "the regulation of the marketplace" as a preferred alternative, whereby if one firm were to provide poor or otherwise undesirable services, they could naturally lose users to rival firms that did better. The technological landscape is littered with such corpses, victims of their own marketplace vulnerabilities. Hewlett-Packard, once a byword for dominance in scientific and engineering hardware, as well as high-risk research, rapidly fell amid the explosion of the low-cost PC clone business in the 1990s. Kodak, IBM, Sun Microsystems, AOL, Yahoo, Blackberry, MySpace: each enjoyed seemingly unassailable product position, until the point where it became outflanked in unpredictable ways. More subtly, when ride-hailing frontrunner Uber's business culture was assailed by aggrieved users and former employees in the 2017 "#deleteuber" social media campaign, the firm saw a four percentage point market-share loss in just a few months to smaller rival Lyft (which described itself to customers as a "better boyfriend").[30] After dating app Tinder's vice president of marketing Whitney Wolf left the company amid public recriminations of ethical violations in 2014, her own start-up dating app, Bumble (positioned as more friendly to female users), exploded in growth and rapidly came to rank third place in overall downloads. In tech, markets adjust.

Finally, and more generally, is concern about unintended consequences. Those brought up in Silicon Valley have an innate sense that while the incredibly innovative technologies that were developed and popularized there were not necessarily done without government help, they certainly benefited from being largely "left alone" to do business, and to repeatedly change how to do business, amid an ever-shifting technological landscape. The result of that has been a mix of good and bad, as with every industry; but the business and consumer products that arose from that environment, products that are used around the world every day to great utility and at low cost, make an argument that the good has been overwhelming. There is a fear then that a jump to government intervention could break that system, either by actually serving to better entrench the positions of those tech incumbents through regulatory capture, or by weakening the overall ecosystem's attractiveness for investment.

Our discussants furthermore observed how the reach of attractive internet technologies to willing users has proved longer than the reach of any one nation's domestic regulatory arm. The American parents of smartphone-wielding middle schoolers may be concerned about the content of their Facebook news feeds, or about Snapchat advertising, and may press regulators to do something about it—but what of the China-based, adolescent-targeting short video app TikTok, which grew to five hundred million users worldwide, including one hundred million US downloads, within two years of its 2017 release? Given the Chinese government's internet sovereignty model, this was probably not the sort of trade that Americans with real concerns on privacy or freedom of expression have in mind. If we are banking on new technologies to enable broad productivity gains in an emerging economy, then we should at every step consider the social costs of limiting that against any expected benefit. Monitoring and careful deliberation are in order.

For more:

- "Unlocking the Power of Technology for Better Governance" by Jeb Bush
- "The Promise of Government" by Amanda Daflos
- "The Commercialization of Decision Making: Toward a Regulatory Framework to Address Machine Bias over the Internet" by Dipayan Ghosh
- "Governance in a World Beyond the News Cycle" by Karen Tumulty

*https://www.hoover.org/publications/governance-emerging-new-world/fall-series-issue-819*

### From "Unlocking the Power of Technology for Better Governance"

*by Jeb Bush*

For technology to reach its full potential, and for government to reliably harness it for the good of the public, we need to transform government itself. Our twentieth-century government—bureaucratic, inefficient, overly centralized, and captured by special interests—is not positioned to make the best of the technological advances that have taken place despite it. Our complacent monopoly of a government needs to be brought into the twenty-first century. We need to cut regulations, make civil servants accountable, fix broken public procurement processes, and, for the sake of accountability, experimentation, and local autonomy, return power to state and local governments. . . .

The American private sector sees a constant hiring and firing of its workforce as managers seek to acquire and retain high-performing employees. A 2016 study by the Department of Labor showed that approximately 1.3 percent of private-sector employees are discharged from their jobs each year. This stands in stark contrast to the public sector, in which job security is treated as a right. Public-sector employees were discharged at a rate of approximately 0.4 percent, which is less than a third of the rate in the private sector. . . .

The problem is not tied to any specific geographic region or government department. In Chicago, it takes eighty-four steps to fire a park worker for incompetence. In New Hampshire, a Department of Labor employee brought her case to the Personnel Appeals Board to protest a formal warning she was given for sleeping at her desk. In her complaint, she alleged that "she had not been given sufficient time since her first warning to correct her problem of sleeping at work." Stories abound of ineffective teachers who are virtually immune from being fired because of the strength of teacher tenure in public schools. In New York City, hundreds of teachers who have been removed from the classroom for misconduct or incompetence are paid to sit in an empty room for six hours a day while the city painstakingly processes claims and appeals related to their incompetence.

CHAPTER 11

# Demography and Migration

As we have seen, the world's population is being reordered. By 2060, without immigration, the working-age populations of Europe, South Korea, Japan, and the United States would shrink by a collective 132 million people, and Germany and Japan would have more people over the age of seventy than under the age of twenty. Over that same period, sub-Saharan Africa's working-age population will increase by nearly one billion. That region, plus other developing countries including India, Pakistan, Egypt, Sudan, Indonesia, Iraq, the Philippines, and Afghanistan, will account for the vast majority of the world's new working-age men and women.[31]

This startling dichotomy is the product of what professors Jack Goldstone and Larry Diamond have described as "twin demographic challenges": a demographic implosion in the advanced industrial democracies due to persistently low fertility rates and a demographic explosion in some developing regions due to persistently high fertility rates. So the developed democracies, for the most part, have aging populations and shrinking workforces, while those of sub-Saharan Africa and the developing nations listed above have rapidly expanding youth and working-age populations. In other words, the balance of the world's population is shifting to its poorest areas, and the bulk of the world's new babies will be born into countries with weak governing institutions—as we explored earlier, sub-Saharan Africa, parts of the Middle East, and the Northern Triangle of Central America suffer from poor governance and are disproportionately affected by a deteriorating climate. The sources and implications of this "toxic brew," as Goldstone and Diamond call it, are myriad. But let us focus on a few salient issues and chiefly on their implications for US policy.

What will happen to the advanced democracies as they age? With fewer young people and an older workforce, they will see shortfalls in the labor force, particularly blue-collar work; an unbalanced population, with more elderly and fewer working-age people to support them; and productivity deceleration, as the productivity of workers plateaus as they get into middle age. We should encourage formal work among educated older workers and women to make the most of that potential labor force. And we should aim for new technologies to catalyze the productivity growth needed to mitigate against the lost workers, but as we learned in our discussion of the future of the US economy, most tasks still will require humans. So developed democracies will also have to lean on immigration to sustain their workforces.

Fortunately, there will likely be many migrants looking for new homes. Unfortunately, the outlook is not so simple. Migrant flows likely will be unsteady, more boom and bust than steady stream, and, as Jack Goldstone has described it, immigration flows are like river basin flows: both floods and droughts can be ruinous.

Why do we predict floods of people into the advanced democracies? Because governance is poor and, in some cases, getting worse in the regions that will experience a population explosion. International assessments of governance quality show negative trends for political rights, civil liberties, and the rule of law in sub-Saharan Africa. The Middle East and North Africa have even worse governance, with particularly predatory governments in Egypt, Syria, and elsewhere. Central America's troubled Northern Triangle lacks almost any rule of law, has stunningly low per-capita income and government revenues, and is wracked by violence. Moreover, in each of these regions, fertility remains high; good governance can create the conditions that reduce fertility—such as better education (particularly for women and girls), employment opportunities, and child and health care—but weak governments rarely do. Add climate change, with the attendant extreme weather events and oscillations and rising sea levels, and we have Goldstone and Diamond's "toxic brew."

Bad governance encourages people to emigrate in search of opportunity and increases the likelihood of conflict or civil war, which generate refugee crises. The Western world saw the results of such crises in the past decade: the surge of migrants to the United States from Central America overwhelmed our unprepared immigration system, while Europe was disrupted by the Syrian refugee crisis triggered by state failure, climate change, and conflict. The political and social costs of these experiences are evident today; waves of anti-immigration sentiment tend to succeed waves of unmanaged immigration. And we can reasonably expect to face more such crises.

It is, therefore, in the direct interest of advanced democracies to help improve the quality of governance, reduce fertility, create economic opportunity, and handle the effects of climate change in these developing regions. Of course, governance, fertility, and economic opportunity are intertwined. It is hard for countries to generate sustained economic growth and widespread opportunity when burdened with poor, particularly predatory, governance. It is also necessary to improve access to and the quality of education, especially for girls, and economic opportunities, especially for women, as well as health and child care, family planning services, and the like. As we have mentioned, improving secondary education for women in sub-Saharan Africa is the crucial factor to reducing fertility there. The example of Bangladesh (see chapter 6) shows that through taking a few key steps, governance can be "good enough" to quickly and dramatically lower fertility, particularly with civil society on board.

The advanced democracies should endeavor to increase the "supply" of good governance through the carrot of foreign aid and investment and the stick of international pressure. They should also encourage in-country, bottom-up "demand" for good governance, which will be critical to the establishment of long-term stability and growth. To address the supply side, these democracies can focus on improving the rule of law by offering help in combating crime, promoting transparency and countercorruption efforts, and supporting political reforms. The United States, as we have described, could reduce violence in Central America by abandoning its misguided "war on drugs" and developing a domestic drug control policy that instead puts the focus on discouraging use. Developed states could also support nongovernmental organizations and indigenous civil society, encouraging the kind of homegrown demand for more opportunities and better governance that would form the foundation of long-term economic growth.

During our roundtable discussions, participants noted that younger populations tend to be eager to mobilize and engage in political activism but grow more stable and democratic as they age. So history would suggest that if the developing states with rapidly growing populations can lower their fertility rates and increase the median age of their societies, they will move in the right direction on social and governance maturity, too. There are, however, notable exceptions to that rule, including China. The People's Republic of China has a relatively high and rapidly rising median age of more than thirty-five and is headed toward achieving the status of a fully "developed" state, but, defying the trend, it remains autocratic. It has progressed through the cycle of reducing fertility and capitalizing on its demographic dividend, and now faces demographic challenges similar to or even worse than those of the advanced democracies.

We can, though, take a few lessons from the example of China. Even before the introduction of its one-child policy, instituted from 1980 to 2015 in response to worries over the country's ability to provide for a growing population, its fertility rate declined sharply from more than six births per woman to under three in just over a decade. However, with its often harsh local enforcement of birth limits, China ran the clock forward on a ticking demographic time bomb. Gender selection led to a severe imbalance of boys over girls. Families shrank, creating increased burdens on nonfamily means of support for the elderly. And the overall working-age population is now shrinking, too, with negative implications for economic growth and national vitality. Today, the Chinese government actually finds itself encouraging multiple children, to the resistance of many of citizens who have grown used to investing all family resources into a single offspring. So we should be careful what we wish for when talk about fertility reductions, but we can also take comfort in knowing that as an immigrant-friendly nation, the United States has the means to ameliorate its own demographic realities, a privilege China does not enjoy. China's remarkable progression reminds us that societies will not inevitably follow the US lead as they age and develop. Despite its self-inflicted ills, China remains a compelling example for the governments of young countries that may be tempted toward an authoritarian-backed narrative of development, given its undeniable progress in bringing millions out of poverty, and the Chinese government is advertising itself as such. And in so doing, it makes a compelling case for the United States and advanced democracies to forcefully argue the merits of free and open societies.

While looking abroad, the wealthy democracies must also put their own houses in order. They will face ever-larger surges of people seeking opportunity or refuge inside their borders. They must, of course, control their own borders—such is the right and the responsibility of a sovereign state—but they also should recognize that as their workforces shrink, they need immigration to sustain their economies and support the elderly. And they often have a moral responsibility to absorb deserving refugees and asylum seekers. As we have learned in recent years, leaders will have to balance the social and political costs with the benefits of immigration, and they should establish policies to welcome, absorb, and manage migration in a more orderly and valuable way. Disorder breeds distrust in immigration, and distrust leads to instability.

Our walk around the world reveals the stark demographic contrast between aging, wealthier nations and rapidly growing, youthful developing states. We surveyed the astonishing population growth in sub-Saharan Africa and the hollowing out of European societies. We saw China's demographic headwinds and the rapid aging of Japan's society. But what of the United States: how does it fit into this picture of shrinking advanced economies and booming poor ones?

The United States enjoys a more favorable demographic outlook than Russia, China, and America's peers, largely thanks to immigration. Like the European democracies, Japan, and South Korea, the United States is aging, but is doing so more slowly. And American society will continue to grow through midcentury as people from around the world emigrate to our shores. The United Nations projects that the US population will grow by roughly fifty million between now and 2050, reaching almost 380 million.[32] Sustained immigration will drive that growth, and immigrants—and their future children—will also spur the working-age population (defined here as ages fifteen to sixty-four) to increase by over 16.5 million people in that time frame. By contrast, if immigration were to cease in 2020, the American working-age population would shrink by 13 million people (over 6 percent) in the ensuing thirty years.

In fact, immigration is becoming the dominant source of domestic population growth overall. In 2017, the US population grew by 2.4 million people, according to the US Census Bureau. Net international migration contributed one million of those people, and the rest can be attributed to "natural increase." That is, there were 1.4 million more births than deaths in the United States in 2017. That balance will shift in the next two decades. The Census Bureau predicts births will outnumber deaths by only half a million in 2040, while total immigration will inch up to 1.1 million.[33]

Why will this shift occur? Large, older generations, namely the baby boomers, are approaching the upper age brackets, and Americans do not have that many children. The World Bank puts the US fertility rate at 1.8 births per woman—higher than that of most rich countries, such as Germany (1.6), South Korea (1.1), or China (1.6), but not high enough to keep society from aging.[34] So, given current immigration levels, the median age of the United States is steadily climbing from roughly 33.0 in 1990 to 38.3 in 2020 and up to almost 43.0 by 2050, at which point the older-than-sixty-five population will outnumber the under-twenty crowd. A decade from now, over one-fifth of the country will be old enough to receive Social Security, and there will be fewer people of working age for every one of them—just 3.4, down from 4.5 today.[35] (That ratio, it is worth noting, suggests we may need to reconsider how we think about the idea of "working age.") And, although life expectancy has plateaued and even declined—down to 78.6 in 2017—we still live longer than we did at any point before 2010.[36] All told, the population distribution of the United States will increasingly resemble a pillar, as opposed to a pyramid. We will not be a youth-heavy society but one distinguished by an even distribution of people across all age groups.

The composition of that pillar deserves further attention. Women will continue to outlive and outnumber men, especially at the oldest ages. Meanwhile, the racial and ethnic makeup of society will change. Non-Hispanic whites will continue to decrease both in absolute numbers and as a portion of society. According to the Pew Research Center, the immigrant share of the total population is nearing a historic high—at 13.6 percent in 2017, just shy of the high of 14.8 in 1890—and foreign-born women give birth at higher rates than US-born women do. So not only will the foreign-born population continue to shift the diversity of society, but their children will as well.[37]

As we have already discussed, while in-migration has increased, internal migration—or mobility—has slowed. A recent study by the Federal Reserve Bank of New York found that only 1.5 percent of people moved to a different state in 2018, half the rate in 1985. Nor are Americans moving more locally; movement within states or even within counties similarly declined. When people do move, they are almost entirely moving to the southern regions of the country, and out of the Northeast.

The general story of US demographics is positive. We have a growing workforce and a population that is younger and aging more slowly than those of other wealthy democracies, as well as those of China and Russia. But much of that is a result of the United States being the world's premier destination for immigrants. We attract the top graduates and those looking for a better future and enjoy a relatively strong demographic outlook as a result.

Since the 2008 recession, there has been much popular and academic discussion of economic inequality in the United States, with the spotlight on "the 1 percent" and billionaires like Jeff Bezos or Mark Zuckerberg. But focusing on the top misses the point. Jeff Bezos helped to create new growth that benefited others, too. Instead, if one is really concerned about Americans' well-being, one should focus on the bottom (How are the least-well-off Americans doing over time, and are they improving?) and on the middle (Are wages of those in the middle of the income distribution going up, down, or sideways?).

And to that question, history would suggest that rapid economic growth rates are among the best ways to benefit Americans at the middle and lower rungs of society. When unemployment rates fall to their lowest levels and labor markets tighten, social groups with generally higher unemployment rates—women, minorities, the elderly, the less educated, or even those with criminal records—are disproportionately pulled into the workforce or are more likely to upgrade from existing jobs. The converse is also true during a downturn: "last hired, first fired."[38] During the depths of the 2008 recession, black employment rates were falling at 5 to 6 percent per year, nearly double the 3 to 4 percent rates for whites. But since 2012, black employment rates have recovered at 3 to 4 percent per year, versus approximately 1 percent for whites. Similarly, unemployment rates for workers with a college degree were 10 percent lower than those for high school dropouts in 2009, but after nearly a decade of continuous economic expansion, that spread had fallen to just over 3 percent. So as a general rule this points to the importance of economic performance as a key enabler of social equality.

In an emerging world, however, our concept of labor markets is expanding. Globalization has set off the freeing of capital flows around the world, allowing for huge cross-border investments and speculation. And it has ushered in the global movement of goods as complex multicountry supply chains have been established to take advantage of beneficial regional attributes and as trade has soared. Outside of elites, however, globalization has not fully unlocked the third leg of the economic stool: labor. Cross-border services—e.g., Philippine call centers fielding US customer inquiries, American banks consulting on European business deals, even international telemedicine—can be considered a form of labor mobility, and the value of trade in services is gradually rising versus that of trade in merchandise, particularly in advanced countries like the United States. But what of the movement of people themselves? We know a strong and dynamic economy pulls in marginalized workers in our own country's labor market—and increasingly we see this now pulling in workers from other countries as well, both within and outside of the legal frameworks in place for that.

Over the course of our project, roundtable discussants considered the history of and continued role for immigration in the United States from this economic perspective.

In 2017, about 258 million people around the world (3.4 percent of the population) lived outside their country of birth. And countries consider multiple factors as they weigh how many immigrants to admit, with what skills, with what status. These include security issues, cultural and ideological concerns, economic interests, and rights. Migration expert Jim Hollifield offered three such frameworks across the unique US immigrant history: the "Massachusetts model," which welcomed immigrants on the basis that they assimilate to the host culture; the "Pennsylvania model," which treated newcomers essentially equally given a baseline respect for local law and basic values; and the "Virginia model," which focused on bringing immigrants for labor.

The United States has applied those models to varying degrees across four major waves of immigration, and the sources of new arrivals have changed as well through those waves. Before 1820, English and Scots dominated—and Africans were forcibly immigrated through slavery. Between 1840 and 1870, economic motivations drove Irish, German, and Scandinavian migrants, including many Catholics. From 1880 to 1914, Chinese migrants went to the western United States for work and to escape upheaval at home, while southern and eastern Europeans went to the East Coast, Midwest, and Southwest. Finally, from the 1970s to the present, migration has included both low-skill and high-skill workers from Mexico, Central America, and Asia. At each step, immigrants reshaped American society and played an increasingly important role in the economy. And each wave drew a reaction from other Americans, some of whom were directly impacted economically by their arrival, and others who may have only considered themselves to be.

The foreign-born population in the United States today has grown almost fivefold since 1970. Discussants described how immigrants are increasingly important for economic growth, providing both labor and human capital. Today's immigrants, for example, are going to the states where the highest economic growth is. High-skilled immigrants such as engineers and scientists, nurses and doctors are a boon to the economy. And importantly, immigration now provides 30 percent of US population growth—without immigration, the US population would have already stagnated and started to decline, like so many other countries in the emerging world.

The most pressing immigration issue today is of course the twelve million people living in breach of US law, half of whom have entered illegally and half of whom have overstayed visas in this country. The lack of enforcement of immigration laws poses challenges to state sovereignty and security, the legal system, and civil society broadly. At the same time, the country cannot deport millions of people who wish to be here, some of whom have already undertaken deep struggle or sacrifice to try to become Americans. So this creates a paradox. To maintain economic competitiveness, the United States should keep its economy open to trade, foreign investment, and immigration, but it also needs to control its borders so as to not undermine the social contract and rule of law. Clearly there is no black-and-white answer to this, and it calls for balance.

Our discussants sketched out what an updated immigration strategy might look like, arguing that immigration policy can be both compassionate and "greedy." Currently, for example, 80 percent of legal immigrants in this country come through the family channel—so-called chain migration—where one legal immigrant is able to easily sponsor other family members abroad, some of them distant, to later immigrate to the United States as well. American public opinion is relatively forgiving of this practice, and it has been credited with helping ensure a social safety net for new immigrants who may otherwise struggle in a foreign society. But it does not necessarily do a good job of meeting the host country's goals in terms of desired skills or attributes, and it also is not fair to other desiring immigrants, who may actually provide a stronger benefit in coming to this country but have no family member to help them.

Canadians deal with this issue through a multifaceted point system that balances these considerations with other interests; approximately 60 percent of Canadian immigrants are now "economic" versus the 30 percent that are "family." The United States could do something similar—perhaps focus the family channel on the nuclear family only—while also thinking about potential members with certain demographics or capabilities. This would be similar to President George W. Bush's idea during his administration of "matching willing workers with willing employers." While experiences then and today demonstrate the political difficulty of getting a sweeping deal through, our discussants were optimistic that with the right leadership, and perhaps by staging reforms as smaller pieces to get the ball rolling, the underlying fundamentals support good prospects for a deal. As one of us has said before about Washington, DC, "Sometimes when you know that you are right about something, but everyone tells you it can't be done, you just keep on with it and eventually the fundamentals will prevail."

And the upshot is that this is not new territory. The history of US immigration has never been clean. Yet we remain an immigrant nation in actuality and in self-image. And while those who watch acrimonious arguments over this playing out daily on cable news may find it a surprising claim, we should not have a crisis of confidence in our ability to handle immigration issues. The fact is that this is difficult, but we in the United States—alongside similar immigrant nations like Canada (with a 22 percent foreign-born population) and Australia (24 percent foreign-born population, not including those from the UK)—are actually the best on the planet at handling this issue because of our unrivaled experience. Globally, migration pressures will grow in the emerging world. Ubiquitous information and communication flows will make information about and interactions with foreign countries and citizens easier. Automation and advanced or additive manufacturing can untangle supply chains and their workers. Changing climates may unmoor people from their existing homes or work, causing them to flee unsuitable areas or seek new opportunities. Military conflicts precipitated by new weaponry may do the same. These flows will need to be managed. More importantly, they will need to be governed as they integrate into the broader flow of the receiving society.

Our existing American diversity is a boon to this integration, as it offers a "golden dome," to quote the iconic mayor of Jerusalem Teddy Kollek, under which new arrivals can find any number of suitable ways to live their lives as Americans. It also means that our governance institutions and procedures were designed from the beginning to encourage counterbalancing expressions of diversity around a common stake in the American creed.

In fact, the United States at its founding was probably the most conscious historical effort to set up a government over diversity. The historical background is instructive. Emerging from the Revolutionary War, the country faced a variety of internal regional challenges and remained more "states" than "united" under the loose Articles of Confederation. At the same time, the populace was heavily composed of immigrants and lacked a strongly hegemonic social default. James Madison and other founders in their contributions to *The Federalist Papers* wondered how the destabilizing tendency of various factions "united and actuated by some common impulse of passion, or of interest" would be managed in the hard-won union.

In his review of the subject, Madison, for example, saw diversity—"a division of the society into different interests and parties"—as "sown in the nature of man." As a part of human nature, it would therefore be impossible to remove diversity's causes, whether through oppression or through consensus. In any case, the young state would not be powerful enough to do so even if it wanted to. Instead, the American answer to managing diversity—as diversity is inevitable—would be more diversity: "The diversity in the faculties of men, from which the rights of property originate, is not less an insuperable obstacle to a uniformity of interests. The protection of these faculties is the first object of government."[39]

This argued for creating a framework to allow as much diversity as possible. Not redistribution, not the seeking of a common denominator, not compensation, but protection of the abilities of its citizens to express diverse interests. Allowing as many diverse interests as possible within a large and expanding union would naturally create a political and social system more robustly resistant to dominance by any single faction or the spread of extremism. Were the state to limit any one interest, however, it risked unleashing another. The development of the Constitution (and Bill of Rights) was therefore the founders' way of distributing power so that this diversity could be recognized. Moreover, over time, and not without misstep, the country has learned how to effectively govern over that diversity.

The very structure of a limited federal government based on checks and balances and the ability of state and local governments to have regional authority over matters closer to home, which affect constituents' lives most directly, allowed the nation as a whole to maintain and represent a diversity of opinion, resilient against domination by any one geographic interest or extremist fad. In addition, still today, these "laboratories of democracy" encourage experimentation in governance, the voluntary formation of ad hoc relationships, and the incentive to improve government performance through a form of regional competition. Of course the protection of religious freedoms was also central to the early identity of the country at a time when the Church of England's monopoly on faith demonstrated the impossibility of forced commonality. As Thomas Jefferson reflected in his own letters, "Divided we stand, united we fall."

The country did not always uphold these values. Women's rights, and later civil rights for blacks, nearly split the nation. Even then, parts of society continued to try to keep the lid on the interests of millions of US citizens for decades. When the pressure that built up eventually boiled over in the 1960s, it served as a stark lesson of the continual failure to effectively govern over this diversity. Though this has been recognized in the years since and was enshrined in the Constitution, it is worth considering the years of opportunity lost for not only the oppressed but also the nation as a whole had the value of this diversity been earlier enabled.

To return to our topic at hand, a diversity mindset can also be argued to have applied to the development of the US economy. Acknowledging that some will be more economically successful than others, and allowing them to personally benefit from the value they create (while at the same time protecting the least well-off in society through a safety net), has meant that Americans have long had the chance to be rewarded for their own risk-taking entrepreneurship. Unlike in some other modern societies, the existence of wealth in the United States—and a shared opportunity to realize it—is generally regarded as beneficial and not something morally corrupt to be appropriated by the state or stamped out through excessive redistribution. This underlying sense of responsibility has preserved a strong incentive for self-betterment across a diverse society. And it will help Americans at all strata of society to find creative ways to take advantage of this century's emerging new technologies for their personal and community benefit.

Jerusalem's Teddy Kollek had this in mind during a visit one of us took there in 1969 to consider the problems we and Israel both shared at the time—bringing people with distinct disadvantages into the labor force. On an unforgettable night, on a tour from party to party around the city, Teddy demonstrated the incredible variety and backgrounds of the people in his city. This was a lesson. He would say, "Jerusalem is a beautiful city, but you have to understand it. Think of a painting. In most paintings, the colors are blended. But in Jerusalem, there are many different categories of people—Jews and Arabs, others too. So you have to think of Jerusalem not as a painting, but as a mosaic. And the job of the mayor is to make the mosaic into a beautiful picture." As mayor, Kollek was someone who understood 1) that there was diversity, and 2) that it needed to be explicitly dealt with to make it work. And it did work.

The United States of course remains a nation of immigrants. The foreign-born share of the population today, about 14 percent, represents on the one hand a rapid increase over recent decades, palpable to most Americans. But from another perspective, it embodies a return to the consistent 10 to 15 percent range that held from the time records were first kept in the nineteenth century up until the Second World War. The postwar immigrant trough reached its nadir of 4.7 percent in 1970—likely the lowest share ever in the history of the United States. So today's situation may feel unfamiliar to a generation of Americans who grew up in that era, but it would not necessarily be something alien to their parents, or their grandparents before them. As an example close to home here in the San Francisco Bay Area, nearly as many residents of this region were born outside the United States as were born in California, and English is the primary language at home of less than half its residents. It may be surprising to learn that native-born Americans make up less than one-third of Silicon Valley tech workers. But somehow it all works. Regional labor productivity now exceeds $200,000 per employee and has grown 50 percent faster than the US average since the turn of the century, and the region's share of US patents granted has doubled. Other countries with less modern experience in this realm are now figuring out how to get there. San Francisco is an example of a city that has its own problems, but that has most notably created a fabric in which anyone can find a place to thrive. More than a third of its residents were born outside the country, and many more than that outside the state. So, when its City Hall recognizes each resident consulate's home national day each year by flying that country's flag and playing its national anthem, it is a way of recognizing that our residents, from their countries, play a contribution to our society.

The going can get rough, and US immigration today has problems, too. But to take the longer viewpoint, this is in our gut. America's ability to incorporate newcomers while maintaining the diversity they brought with them is unrivaled, and that will help us to make the best of—and provide needed global leadership in—what may be an even more chaotic global migration landscape going forward.

We find ourselves at an inflection point, in need of good governance. Yet good governance is in short supply these days. The United States and its European partners, whose leadership shaped the second half of the twentieth century, have seen their institutions come under attack or atrophy. Some democracies, as our Hoover colleague and political scientist Mo Fiorina pointed out in the course of this project, are struggling to adapt to early demographic and technological changes. Meanwhile, authoritarian systems are strengthening their grips on society and increasing in number, with China in particular looking to these technologies to rearrange society and help the Chinese Communist Party implement a modern authoritarian state. And as we've described above, many developing countries have failed to establish stable, effective institutions. Stability everywhere seems precarious.

Against this backdrop, the United States should work with like-minded advanced democracies to address the preeminent challenges of the emerging world. Indeed, the United States enjoys an almost unique position. Unlike the vast majority of developed countries, we are projected to experience population growth in the coming years, thanks to both (so far) moderate fertility rates and immigration. As a nation of immigrants, we have a firm foundation to build on, an ability to welcome, absorb, and assimilate peoples from all over the world, and a common civil creed around which to unite a diverse society. We can set the example for those countries less experienced in taking in and governing over diversity.

The United States also possesses unmatched economic capacity and political authority. We remain the preeminent economic and military power, and we enjoy the general support of the other advanced democracies. That coalition of the willing, led by the United States, would have the capacity to balance against authoritarian influence, motivate international engagement, and rally the world's wealthiest states around the common concerns outlined above. And as we discuss next, the United States, in many ways still among the leaders in addressing climate change through its market- and regulatory-driven pursuit of natural gas, efficiency gains, advanced technology development, and laboratories of climate policy in the states, can lead on that account as well.

Of course, the United States has its own internal drivers of instability and discontent. It is no secret that trust in and satisfaction with government has waned, and that democracies generally are struggling to adapt to demographic and technological change. The costs of that democratic distemper, as Fiorina calls it, include civil disorder, the breakdown of social discipline, the debility of elected leaders, the alienation of citizens, and the loss of trust in nearly all institutions of civic and political life, aside from the military. But we have seen this movie before, and it was far more violent; the era that we think of as the 1970s—technically beginning with the late 1960s—

saw a crisis of democracy in both the United States and abroad, notably in the United Kingdom. Then, as now, the crisis was driven in large part by the sense that government had failed, while engaging in misbegotten foreign wars, facing youth protests and new civil rights challenges, and presiding over weak economies. It was driven as well by social and technological phenomena—migration and the ubiquity of media sound bites, for example. In some ways the crises of both the 1970s and today reflect the fundamental truth that democratic systems articulate interests well but aggregate them poorly. However, the distemper of the 1970s gave way to a generation of prosperity and strength. And it did so largely because electorates chose leaders who committed themselves to solving hard problems and earned long-lasting, stable majorities as a result.

What made those leaders successful? From where we saw it, they trusted in the proven successes of classical, orthodox policies. The Reagan administration committed to peace through strength and responded to the stagflation of the 1970s with sound monetary policy to get the money supply under control, weathering short-term political backlash in doing so. It prioritized its political capital in remarkably effective ways. Despite a challenging world, it resisted the reactionary impulse. We can learn from that example. New problems—shifting demographics and new technologies—demand firm responses but not necessarily new ones, and when confronted with the threat of global instability, the leadership of the United States is indispensable.

For more:

- "How Will Demographic Transformations Affect Democracy in the Coming Decades?" by Jack A. Goldstone and Larry Diamond
- "The Democratic Distemper" by Morris P. Fiorina
- "On 'Forces of History': Easy as One-Two-Three? Not Exactly" by Charles Hill
- "The Migration Challenge" by James F. Hollifield

*https://www.hoover.org/publications/governance-emerging-new-world/spring-series-issue-719*

## From "How Will Demographic Transformations Affect Democracy in the Coming Decades?"
*by Jack A. Goldstone and Larry Diamond*

The biggest threat to democracy around the world, in terms of demographics, is the combination of fast-growing and youthful populations with poor governance, mainly in Africa, the Middle East, and South and Southeast Asia. This combination leads to higher risks of civil conflict in these regions, which threatens their own chances for stable democracy. It also undermines resiliency in the face of political or climate disasters, which, as population swells, is likely to produce ever larger waves of refugees heading to Europe or the United States. Today, a civil war in Syria (population 19 million today, 30 million by 2040) has propelled a million refugees to Europe. Imagine if the next civil war is in Egypt (100 million today, 137 million by 2040), or Ethiopia (113 million today, 166 million by 2040), or Nigeria (more than 200 million today, more than 300 million by 2040). Another huge eruption of asylum seekers would likely strengthen the populist, anti-immigrant, illiberal parties in Europe, and the similar movement in the United States, that have been gaining strength over the last decade.

There is, unfortunately, no silver bullet for providing good governance in developing nations. Rigorous research shows that democracies, in general, promote economic growth better than dictatorships do. But that is an average result, and the effect is modest—a long-run relative gain in income per head of only 20 percent. Moreover, countries rarely transition from autocracy to full democracy; most transition first to partial democracy, which is a highly unstable and conflict-prone stage of development. Other things being equal, liberal democracies have the advantage of providing more accountable governance, strengthening civil society, and encouraging individual initiative. But regardless of regime type, a government that enforces property rights; invests wisely in education, infrastructure, and industrial capacity; provides inclusive economic growth and good public services; and has decent macroeconomic management will be more resilient to political and climate-induced crises. It is thus in the direct interest of the advanced democracies, if they wish to avoid intermittent crises of waves of refugees, to address the problems of predatory and corrupt governance in developing nations and seek to encourage the good governance practices that reduce the risk of unmanageable crises.

CHAPTER 12

# Health and the Changing Environment

Human societies have generally made great progress over the course of history in the mastery of their surrounding environments, climates, and biomes. And across a variety of measures, the experience of the United States is emblematic of this, with significant reductions in air and water pollution, in weather-related mortality, in malnutrition, and in the burden of disease. Progress has been spurred by a combination of technology, markets, and governance. As described by scholars including the Massachusetts Institute of Technology's Andrew McAfee, capitalist competition has driven a new form of dematerialization across the United States and other developed economies, with oftentimes not just relative but even absolute reductions in raw material inputs to our economy—things like copper, steel, stone, fertilizer, even paper—and the waste and pollution that result from their use. Frequently, difficult social and regulatory choices over the past half century, enabled by technological innovation and ongoing incentives for investments, have allowed this country to stay one step ahead of the variety of environmental and health risks it faces.

It is perhaps indicative of the consistency of this progress, though, especially since the 1970s, that the continued fragility of our modern relationship with the environment goes unappreciated. But with anthropogenic climate change and other human-ecosystem impacts, some of these natural risks are actually set to grow faster, stronger, or in newly unpredictable ways. Our exposure to those environmental damages may also increase in the United States, as an increasingly prosperous society with more built assets in harm's way, to say nothing of encompassing an aging population that could be more impacted than younger societies by infectious disease, extreme weather, or other ecological events. Even today, despite major successes in recent decades, a conservative estimate would likely still ascribe more than

10,000 American deaths each year to pollution.[40] The environment is not a solved problem. Businesses and other individual actors throughout the economy will have to continue their relentless march toward innovation and efficiency, broadly defined. And governments will have to find new ways to stay ahead of these changes and protect their populations from threats whose mitigations may now be taken for granted, after years of the worst environmental and biological effects having been successfully staved off.

Our roundtable discussants identified a number of such underappreciated environmental risks going forward. And while many of their observations were not scientifically novel, they were put in terms that make them accessible to citizens and politicians who tend to prioritize today's policy problems before tomorrow's. Moreover, our discussants proposed to use new technologies, particularly in diagnostics, to facilitate implementation of traditional approaches to countering many of these increasing risks.

## Pandemics

Stanford historian Kathryn Olivarius has described the development of social standing in antebellum-era New Orleans, perched on the mosquito-laden mouth of the Mississippi River, as a hierarchy of immunocapital: new white arrivals to the economic center for cotton would be encouraged in newspaper advertisements to "Get acclimated!" to the scourge of yellow fever so that their ensuing immunity to the disease would permit them to become credible members of society in a city that lost almost a tenth of its population to disease in a given year. Despite a 50 percent risk of dying in the process, acclimation was required for citizens to take out bank loans, start a business, sign a lease, or get married.[41]

American governance, scientific knowledge, and private innovation have of course helped public health make great strides since then. Around 1900, an estimated ten thousand or more Americans (from the much smaller US population at that time) were dying each year from malaria, and at least that many from smallpox. But domestic transmission of both pathogens was essentially eliminated by 1950.[42] American life expectancy rose from fifty years in 1900 to seventy-nine today—and over the same period, the survival rate for children up to age five grew from just 82 percent to 96 percent in 1950, and over 99 percent today.[43] And since as late as 1980, US infectious disease deaths of all types have fallen by nearly one-fifth.[44]

But infectious diseases, which had declined during the twentieth century, are making a comeback, and our discussants argued that pandemics once again pose a major threat to humanity. The debilitating COVID-19 pandemic of 2020 is a clear example of this; but in the emerging new world, the spread of the SARS-CoV-2 virus should be seen as an expected event, not a one-off black swan. Part of this is due to growing global human contact. Mobility is increasing everywhere alongside the growing affordability of longer-distance travel. Since the 1980s anthropologists have posited the concept of a "travel-time budget": that on average individuals in societies spend approximately the same amount of time each day, roughly one hour, devoted to traveling to and from work.[45] As technologies improve, people become richer, and the price of mobility falls, resulting in people choosing to travel longer distances at higher average speeds for economic activity. This improves the efficiency and productivity of labor markets, as workers of all skills gain better access to the employers that are best suited to them across a broader geographic area. But it also creates new pathways for the spread of infectious disease in and out from urban centers, and even across continents.

Global urbanization rates are also rapidly increasing, especially in the developing world, due in part to demographic shifts. In parallel, increased immigration to developed countries in temperate regions from source countries across the global tropics—and ongoing personal interactions with family at home—has increased the prevalence of tropical sickness in northern hospitals. As one of our contributors, emergency room physician and Stanford professor Milana Boukhman Trounce put it, ubiquitous global air travel has become an infectious disease "supervector."

Meanwhile, human populations are also changing the nature of their contacts with animal disease pools or vectors, due in part to climate change. Shifting precipitation and temperature patterns, for example, are redistributing the spatial and temporal incidence of disease-spreading mosquitoes—which perhaps will ultimately result in a lower incidence in some areas but a higher one in others, including the introduction of tropical diseases into the United States. As another example, African Ebola outbreaks are not new, but the 2014–2015 Ebola outbreak ended up affecting far more people than did previous outbreaks, spreading across the entire continent rather than within a single village, with even worldwide impacts due to international travel. Similar dynamics now apply to the spread of predictable outbreaks, such as the annual flu, as well as other infectious organisms.

Beyond these natural drivers, the future also holds the risk of disease outbreaks from accidental lab release (with now hundreds of infectious disease research labs around the world, growing rapidly and with varied levels of local oversight) or even deliberate means as bioweapons or terrorism. The rapidly declining cost of genetic engineering will exacerbate both risks as it becomes more attractive to engage in pathogen manipulation for purposes good or ill. Humans ourselves are at risk from this, but so are our livestock and other agricultural economies.

What can governments do about this threat to mitigate new risks and extend the significant progress in public health that has been made over the past century? Our discussants considered what has worked in the past, what has not been effective, and the potential for new approaches.

Prevention could take two forms. One promising route would be mitigating the newly evolving activity of vectors. As described above, mosquito population control has been practiced for over one hundred years to immense human benefit. Going forward, newly developed genetic engineering "gene drives" offer numerous new options, with varying trade-offs between efficacy on the one hand and ethical concerns or the risk of unintended consequences on the other; such techniques can reduce vector (e.g., mosquito) populations, limit vector reproduction, or inhibit the ability of the vectors to transmit a pathogen. In many cases the biology to do so has already been developed. In weighing the governance framework for deployment of such technologies, we would urge that their potential costs and benefits be considered not against a standard of perfection but against a status quo in which the human burden of infectious disease is already large—and set to grow in a changing world.

Preventing new infection through new vaccines is harder. Humans and the animals we value are potentially vulnerable to thousands of strains of viruses, which can mutate rapidly. Developing vaccines or drugs for all of them, which can easily cost over $1 billion each, is not feasible. Even today's widely administered flu vaccine, for example, which is developed ahead of each flu season based on predictions of the form the virus will take, is of uneven effectiveness because the virus mutates rapidly. And therapeutic antiviral drugs like Tamiflu have been shown to be of little value in easing virus symptoms, apart from in patients with already compromised immune systems.

But could new technologies help speed the development of disease treatments once an outbreak has already begun? Traditional vaccines may take a few years to develop, but our panel discussants advised that even a large infectious disease outbreak could run its course in twelve to eighteen months—before conventional vaccines would ever be ready.

One strategy that could help is more rapid detection and early characterization of an outbreak. Renowned biotechnologist Stephen Quake described how biology has become an information science—that is, one driven by data—enabled by low-cost and scalable cloud-based computing and data storage. Meanwhile, the cost of sequencing DNA has dropped by orders of magnitude over just the past ten to fifteen years, at rates comparable to exponential "Moore's law" microchip improvements from the 1980s through the 2000s that helped fuel the explosive growth of the computer industry. Together, these two advances mean that biologists have rapid access to the genomes of many organisms, including human and animal pathogens at an early stage of a pandemic. In conjunction with using the right tools, this can cut months out of the detection process.

These same technologies can help develop rapid therapies. Quake described, for example, how his team was able to take DNA samples from a population at an early stage of a novel dengue outbreak and rapidly sequence them against a cloud database of known DNA patterns to look for new strands—including disease antibodies that had been generated through the immune responses of some of the infected individuals. Reproducing those highly targeted antibodies and delivering them to other infected or at-risk individuals offers a new path to quick, almost "in the field" treatments.

Some challenges are less amenable to technological solutions. A large pandemic would of course have direct effects in compromising the health of many citizens. But even a smaller infectious disease outbreak with high mortality rates could be debilitating given the somewhat prosaic difficulty in providing surge capacity—additional hospital beds, clean rooms, water supplies, etc.—in the health-care system. Our discussants described from their firsthand experience how even a very large regional trauma facility such as the Stanford Hospital could host, at most, ten Ebola patients at one time—and that even that would require the complete shutdown of the hospital's cardiac wards. Just one year after our roundtable, the transformation of our local fairgrounds and hotels into emergency triage centers for COVID-19 infections, staffed by the National Guard and retired volunteer nurses, gave a sense of what an even more deadly outbreak could bring.

While the development of rapid and accurate diagnostic tests such as what Quake described—at-home tests for Ebola, for example—would be crucial, the historically effective solution in such circumstances would be public-health measures: isolation (of those with the disease) and quarantine (of those exposed). Local governments at the county or city level would be responsible for creating and enforcing bans on public gatherings, school and business closures, and strict home quarantine and isolation policies. Doing so has always been difficult, with one panelist observing that during the 2003 SARS outbreak, "even Canadians in Vancouver didn't want to comply [with quarantines]."

New, quick, and accurate diagnostic tests could enable effective implementation of isolation and quarantine. Panelists additionally speculated whether new information and communication or logistics technologies could facilitate such difficult-to-enforce measures through location tracking, telemedicine, or the automated delivery of food, water, and supplies. Overall the panelists argued that "the public sector is not sufficiently preparing for this." While local public-health departments have protocols, do drills, and receive guidance from the federal Centers for Disease Control and Prevention (CDC), "there are too many cooks in the kitchen." A large-scale disease outbreak is not just a medical event but also one of public safety and security: "It's chaos every time, and it is always reactive." Looking at how warehousing and logistics technologies are developing, panelists considered how the US private sector might end up delivering many needed services in such an outbreak, and the market incentives and coordination that governments could consider to help enable that.

An early retrospective of the US governance and civil society response to the COVID-19 pandemic of 2020 changes little from these warnings. It has simply made them clearer. We understand that political realities make proactivity—fully preparing for every small-chance, large-impact risk ahead of time—infeasible. No one celebrates the mitigation of the outcome that never occurred. One could say that the American way of dealing with the problem of prioritization is instead to react, swiftly and effectively, to the reality that has been made present to everyone. The whole-of-society approach to the 9/11 attacks exemplifies this sort of mobilization. But the COVID-19 pandemic calls that approach into question for the sorts of risks we see coming over the hinge of history: though small circles of experts knew exactly how the problems of a pandemic were likely to unfold, their technical foresight described a scenario so foreign to many of our institutions that once it actually happened, attempts to respond to an unrolling crisis fell flat. It took too long for them to internalize the novel nature of the crisis they faced. This was true across many organs of both government and business. And it highlights the need not just to better prepare for pandemics but to better prepare to react to anything, and to effectively function while under duress. In a changing world, we are not innate masters of our universe. What we have now is the result of battles and reconstructive efforts past. So we too are obligated to build again.

## Climate Change and Environmental Pollutants

A combination of technology and policy within a market-incentive framework has historically led to reductions in many local or regional environmental pollutants in the United States—including ozone, particulate matter, carbon monoxide, sulfur dioxide, nitrogen dioxide, lead, benzene, and mercury. Our panelists asked though how much of a human health threat still remained from these local pollutants, particularly in the form of poor air quality, and if any of that progress on conventional pollutants might be reversed through their interactions with the effects of concurrent, carbon dioxide–driven, global climate change. This is an important question, because as important as global climate change itself is, the impacts on livelihoods and human health from local pollutants are actually far more individually damaging.

For example, the detrimental impacts of small particulate matter remain severe in the developing world. While the pollution of Beijing and New Delhi are extreme examples, 92 percent of the world's population lives in areas where air quality does not meet World Health Organization standards. Particulate matter can be inhaled deep into the lungs, bringing hazardous substances into the bloodstream. Exposure can have economy-wide impacts, with a range of estimates in China attributing annual GDP growth rate losses of between 1 and 4 percentage points to health and childhood developmental costs associated with air pollution. Meanwhile, indoor cooking with wood or dung, a method relied upon by more than a billion people in Africa and South Asia without access to cleaner commercial fuels, produces smoke with particulate levels one hundred times higher than acceptable levels and results in millions of deaths annually.

Even in the United States, the changing risk of and exposure to regional drought (and associated particulate matter dust and pollen), floods (mold), and wildfire (smoke and other toxic particles carried with it) under changing regional climate regimes may put the respiratory health of the public at risk. Going forward, discussants argued for increased use of controlled burns, for example, which burn less intensely than uncontrolled wildfires and therefore result in fewer health impacts for those exposed to resulting smoke. Rising temperatures may increase Americans' environmental exposure, broadly defined: a longer and more intense allergy season, the spread of insect-borne diseases, more frequent and dangerous heat waves, and heavier rainstorms. One emerging area of policy attention has been ingestion of mercury and microplastics through seafood consumption.

There is reason for optimism that we can successfully handle these risks. In the United States since the 1970s we have seen substantial, sustained reductions in per-capita emissions of air pollutants such as sulfur dioxide, nitrogen oxides, and chlorofluorocarbons, as well as reductions in exposure to toxic materials such as asbestos and lead—despite overall economic activity quadrupling over that same period. Today's per-capita energy-related sulfur dioxide emissions in the United States, for example, are about 85 percent below levels of just twenty-five years ago, while nitrogen oxide emission intensity has fallen 60 percent.[46] At that time, particulate air pollution from US coal-fired power plants alone was thought to contribute to thirty thousand American deaths annually; today, with market-driven fuel switching to natural gas and improved (and often government-mandated) emissions controls, that number has fallen by 90 percent.[47]

In addition to conventional pollution mitigation measures, our panelists suggested, other emerging technologies might help reduce these public-health risks going forward. Even in developed countries, the elderly remain at risk from heat waves, and heat is already the largest weather-related killer in the United States; low-cost wearable devices that automatically report on cumulative heat exposure, hydration, and body response could help in an aging US society. (While deaths from extreme temperatures are difficult to measure, studies suggest that extreme heat deaths have already gradually declined in a variety of US cities over the course of the twentieth century, and this progress could be further extended.) Emerging technologies over the last decade now show promise for using the same sort of immunoglobulin E antibodies that cause allergies—which already affect sixty million Americans at a cost of $20 billion per year in lost work days—to actually create effective therapies for them. And going forward, advancements in regenerative medicine may also be able to play a role in mitigating the effects of pollution-related maladies such as lung disease. Our discussants noted that where markets for such services or therapies exist, US industry is already moving rapidly to use these technologies to develop valuable new consumer products. They should be encouraged.

## Ecosystem Services

Stanford biologist Lucy Shapiro and physicist Harley McAdams observed in their investigations for this project that even optimistic projections of reductions in greenhouse-gas emissions over the next century anticipate significant climatic changes, including potential disruptions to valuable ecosystem services such as environmental pollution filtration, groundwater, seasonal precipitation storage, pollination, biodiversity, agricultural productivity, and other food chains. Since before the founding of this country, the United States has struggled against weather and environmental threats, sometimes at great cost—consider the plight of early inhabitants who regularly faced flood and drought, hurricanes and blizzards, or infectious disease and waterborne illness. Today's climate change is perhaps unique though in the number of ways it will affect various aspects of human prosperity. It is also unique in that the development of modern climate science tools helps us foresee these changes with some degree of reliability. From a policy perspective, then, as societies strengthen their efforts to reduce warming, they should also be planning for—and budgeting for the costs of—how to ameliorate the expected human impacts of these damages.

Consequences we are already seeing include population displacements due to rising relative sea levels alongside coastal subsidence (when the ground surface itself sinks, usually due to overdrafting of groundwater resources); extreme fires and storms, causing repeat economic damages; acidification and warming of ocean waters, leading to the decimation of coral reefs; changes in food production; and accelerated extinction of species.

Consider tropical coral reefs, about half of which, according to our panelists, have experienced heat stress or bleaching, with adverse consequences for fish and the other marine life they shelter and support. Recently, scientists have discovered how some corals have naturally adapted to warm water and acidification, and that these populations might be used to help seed new growth in damaged areas through transplantation.[48] While some may find this hands-on approach to be an unsatisfying concept, it echoes the history of human cultivation of the terrestrial plant biome through intense forestry or cultivation practices.

Another example is storms. Extreme precipitation events have increased in the United States over the last century. But the incidence of floods in this country has remained essentially flat, and deaths from floods have declined on a per-capita basis.[49] This is due in part to proactive efforts to control runoff and waterways through infrastructure development or complementary management practices. Similarly, despite incidences of extreme weather, tornado and lightning deaths declined on an absolute basis in the second half of the twentieth century, aided by upgraded building codes, detection technology, and emergency response.[50] This hasn't been a cheap course of action, but it has been worthwhile.

Similarly, we may find ourselves in the United States forced to take novel steps to avoid increased economic climate damages. For example, this country has seen enormous successes in improving the productivity of our agricultural system: US corn farmers produced thirty bushels per acre of farmland in 1900 and forty per acre in 1950; today they produce over 150 bushels per acre. Soybeans and wheat have seen similar productivity growth, and each US milk cow today produces 2.5 times more milk than its predecessors did fifty years ago.[51] And in recent years, agricultural biotechnology has been largely focused on extending the gains of the Green Revolution by further improving those yields. With a changing climate, however, the field may now need to shift more toward the development of drought-, heat-, or disease-resistant crop varietals instead, with the associated opportunity costs of doing so.

Or, in parts of the country that will experience drought, new water infrastructure development may be necessary to both store seasonal runoff and improve end-use efficiency, such as in agricultural irrigation, industrial, or residential use. California is one of these places. But the fallout from the state's struggle to work out the economics of undertaking large new capital investments in its hundred-year-old water delivery system, at the same time that consumers try to use less of it due to drought, is daunting: higher rates, for less water. It's a concrete example of how climate damages are already showing up in quotidian affordability concerns, and with political implications.

Day-to-day human activity is likely to be affected too by climate impacts, which can result in a sort of constant tax on normal livelihoods and commerce. Heretofore in modern US history, for example, vector population controls and a generally temperate climate have insulated us from debilitating tropical diseases, with all their ensuing health-care costs and drains on productivity. But with climate change, mosquitoes are moving north. We will need to redouble our diagnostic, treatment, and eradication capabilities for the diseases they carry in order to hold our ground.

Science, engineering, and the modicum of social governance necessary to apply them were ultimately what allowed New Orleans to tame its harsh environment, as described earlier in this chapter, and shed its antebellum moniker as the "Necropolis." But in a leveed city that owes its modern survivability in part to the US Army Corps of Engineers, it seems appropriate to quote General Jim Mattis's observations on the Iraqi insurgency: "In my line of work, the enemy gets a vote." The floods and storms that New Orleans now faces each year are emblematic of our newly resurgent environmental realities. Going forward, maintaining our relationship with our environment—and obtaining the services it provides for us—may simply take more effort.

## The Intersection of Technology, Markets, and Policy in an Emerging World

The concrete nature of the challenges presented in this volume drives home the point that policies to address climate and the environment should focus less on what may or may not happen in the future, or on abstract numerical targets and complex agreements, and more on what can feasibly and responsibly be done today to reduce our risk. Many of the steps we could take today in anticipation of any future risks would start accruing benefits even now given that we already manage or mitigate our environment on many fronts.

This was one of the lessons of what is perhaps the most effective coordinated international environmental strategy in history: the Montreal Protocol, the 1987 treaty promising to control the use of ozone-depleting chemicals across a wide swath of countries. At first the science behind the initiative was uncertain, but it was compelling enough that were the scientists found to be right, the environmental results would be unacceptable. So we initially took obvious, easier policy steps toward addressing the chlorofluorocarbon problem—but we also got American companies started on developing technology to give us better options to address the more difficult aspects of the problem in the future. By the time that the Antarctic ozone hole was observed, those new technologies were close enough to being mature that we could go ahead with a specific and binding agreement that we had confidence in actually fulfilling. We've done this before.

Similarly, one recurring theme in the day's discussions was the growing relationship between climate change and other "conventional" environmental concerns like air, water, and soil pollution. Some of these systems have scientific links. But they are also increasingly linked in the minds of the public and in policy making. Consider key recent federal "climate" regulations such as the 2015 proposed Clean Power Plan, or even some vehicle fuel-economy standards: both measures promised favorable overall social benefits given their costs to consumers and businesses in implementing them. But what drove that favorable economic balance in the push to pass these rules was not actually the expected social value of avoiding some extra degree of global climate change, despite the focus on carbon dioxide. Instead, their value calculus hinged on the much larger expected "co-benefit" of the concurrent reductions in local and regional air pollution like acid rain or smog that would go along with the carbon reductions being targeted. And whereas public polling by Pew Research suggests year after year that climate change remains relatively low on US voters' list of priorities for policy makers in Washington, DC, maintaining an overall healthy environment consistently ranks highly across party affiliations. Globally speaking, contemporary America's natural environment is one of our country's best assets, and we should be looking to maintain that alongside vibrant economic growth.

Some of the problems are not even so complicated. Today, for example, think of the existing nuclear power plants in this country, which provide substantial amounts of clean energy, and some of which are now unexpectedly having to be shut down by their operators because they are losing money in today's low-cost electricity markets. But we know that when those plants get shut down, they are replaced by polluting alternatives.[52] Fixing this doesn't take some aspiration for the future (though we should be investing in even better nuclear power technologies for the future, too). Instead, this is the sort of three-way intersection of technology, markets, and policy that we will increasingly encounter in a changing environment. While these are the sorts of questions that state and federal agencies have always had to respond to, as the natural pace of change gets faster on the other two of those three dynamics, government will need to get faster at identifying and responding to change as well.

Importantly, while some of the challenges described in this volume are new, and others are as old as civilization, our panelists—themselves scientists, doctors, and engineers—remained compelled by the optimism of an American spirit of discovery and innovation to take them on. This was tempered only by their regret in the strained relationship between science and policy in this country, where knowledge and promise are too often selectively ignored or weaponized by both political parties.

They concluded with a plea for the foundational long-term importance to the nation of investing in science and engineering. Markets enable a host of groundbreaking science and health inventions by private entities. In the climate arena, for example, we strongly endorse a carbon dioxide emission tax-and-dividend approach that would price carbon throughout the economy and subtly shift incentives across the spectrum of market players, both producers and consumers, toward cleaner options (then, importantly, returning 100 percent of funds collected through that tax to families in the form of a flat per-capita rebate to minimize fiscal drag). But public funding is still crucial in serving earlier-stage research that doesn't offer good near-term private investment prospects, and for which socially valuable intellectual property can be made more widely available. In a time of newly emerging challenges, our discussants particularly called for a shift in public funding toward riskier but potentially game-changing forms of research and development that get sidelined by conventional funding processes, which tend to emphasize lower-risk, marginal advancements. Where governments have dialed back those risky research ambitions, philanthropy has tried to fill the funding gap, but its reach is more limited. At the same time, discussants sought to rebalance the working relationship between Washington, DC, and the optimism of labs such as theirs around the country. As mentioned above, Stephen Quake reviewed the many groundbreaking research projects being conducted here at Stanford and at partner institutions, as well as in industry. We are fortunate to be at the center of such great work, and the vibrant culture of investigation and discovery here speaks to the value of fostering such efforts. To conclude with the summation from one panelist, addressing the reticence in the public sphere toward riskier long-range research:

> This argument—about if you should forgo basic, and potentially transformational, but risky research projects with no obvious end point and instead just study more predictable, marginal things that you already understand, and simply apply them to a different application—is a strange argument.
>
> If you don't push the boundaries of understanding this world that we are living in, then when we are faced with a complete skewing of the ecosystem of the globe (which we are dealing with now) without new kinds of understandings of how living beings, living organisms can survive changes in their environment—we are being, if not shortsighted, then we are being criminal. It just makes no sense whatsoever.
>
> But somehow we have to do both. We have to continue digging for new ways of understanding the world around us, so that we can mitigate disasters, and we have to cleverly use what we already know about to create ways of dealing with them and the immediate situation. So you have to deal both with the future and with the present, and have enough science and enough people who are making decisions listening to the science, so that we make intelligent choices.
>
> And right now, both the world ecosystem is skewed and our political system is skewed, and these things must come together, or I fear for what's going to happen to life on this earth.

For more:

- "Potential Pandemics" by Milana Boukhman Trounce
- "Governance Challenges to Infrastructure and the Built Environment Posed by Climate Change" by Alice Hill
- "Climate Change and Environmental Pollutants: Translating Research into Sound Policy for Human Health and Well-Being" by Kari Nadeau
- "Global Warming: Causes and Consequences" by Lucy Shapiro and Harley McAdams
- "Health Technology and Climate Change" by Stephen R. Quake

*https://www.hoover.org/publications/governance-emerging-new-world/spring-series-issue-419*

## From "Global Warming: Causes and Consequences"
*by Lucy Shapiro and Harley McAdams*

Currently, the oceans absorb 93 percent of the heat trapped by greenhouse gases in the atmosphere, thus slowing warming of land masses. But the resulting rapid warming of the oceans directly impacts marine life and related food chains. Consider, for example, the coral reefs along over 93,000 miles of coastline rimming the oceans—one of the largest ecosystems on the planet.

A thriving coral reef is comprised of groups of millions of identical tiny polyps a few millimeters wide and a few centimeters long, each with a calcite skeleton. Millions of these tiny stony skeletons accumulate over generations to form the large, hard coral reefs found along tropical shorelines. Many of the coral species obtain most of their nutrients from photosynthetic algae plants called zooxanthellae. When the sea around them warms excessively, the polyps expel the zooxanthellae and the coral becomes completely white—a condition called coral bleaching. Corals can survive bleaching events and restore the zooxanthellae, if conditions normalize quickly enough. But the bleaching events are highly stressful, and the corals will die if occurrence of bleaching events persists. When this happens, only the dead coral skeletons—which can be immense—are left. . . .

Loss of the ocean reef ecosystems could substantially compromise the Earth's ability to sustain the health and well-being of its inhabitants. Fish populations in the coral reefs are the source of food for hundreds of millions of people. Loss of the reefs disrupts the marine food chain, which causes loss of local food supplies, stressed populations, and conflicts over fishing rights. . . .

When a region's climate warms, it may become hospitable to new vectors, which will then inevitably arrive either by expansion from adjacent territories or as accidental hitchhikers in freight shipments or transport vehicles.

For example, in a remarkably short time, human viruses like Zika, dengue, chikungunya, yellow fever, and West Nile have spread into regions of the Caribbean, Latin America, and the United States that until recently had ambient temperatures below that required to support their transmission. In addition, fungal infections of food plants, like the blights infecting Cavendish bananas and cocoa trees, have become a global problem. The rapid spread of global disease caused by changes in atmospheric temperature, ocean temperature, erratic and drenching rains, and floods in one geographic location accompanied by droughts in another location is being facilitated by migration of the vectors, such as mosquitoes, ticks, bats, and rats, that carry the pathogens. Insect vectors are exquisitely sensitive to changes in temperature, and warmer temperatures increase their breeding season and life span. Zika, dengue, chikungunya, and yellow fever viruses soon follow arrival of the common *Aedes aegypti* mosquito and are then transmitted among humans by the female mosquito. Other mosquito species transmit West Nile virus, the malaria parasite, and the parasitic nematode worm that causes the human disfiguring disease lymphatic filariasis (elephantiasis).

Ticks are another rapidly spreading vector. Although most tick species do not harbor pathogens harmful to humans, Lyme disease is caused by a tick-borne bacterial pathogen, *Borrelia burgdorferi*. Until recently, ticks were inhibited over much of North America by cold winters, but with increasing average temperatures and milder winters they are becoming established further north. Lyme disease is now endemic in Canada, so the government has recently established tick surveillance networks.

### From "Health Technology and Climate Change"
*by Stephen R. Quake*

We are living in a time of rapid technological change, and there are many revolutionary technologies that can be put to work solving the health challenges that are arising due to global climate change. To set the stage for some of the discussion to follow, let me outline some of the important technologies I see playing a role. One technology you will see coming up again and again is that of genomics and DNA sequencing. We are living in the genome age, where biology has become an information science, and this transition was driven by the invention of incredibly powerful sequencing machines. These machines are amazing tools that have dropped the cost of sequencing by orders of magnitude and increased the throughput of sequence data, which we can acquire also by orders of magnitude. These trends are comparable to that of Moore's law in the semiconductor world. As a result, we now have genomes of many major organisms—not just humans but also many of the pathogens that afflict our health. We are able to rapidly sequence novel genomes as pathogens mutate and new pathogens emerge. And we are able to analyze human biology with a power that was only dreamed of a few short decades ago.

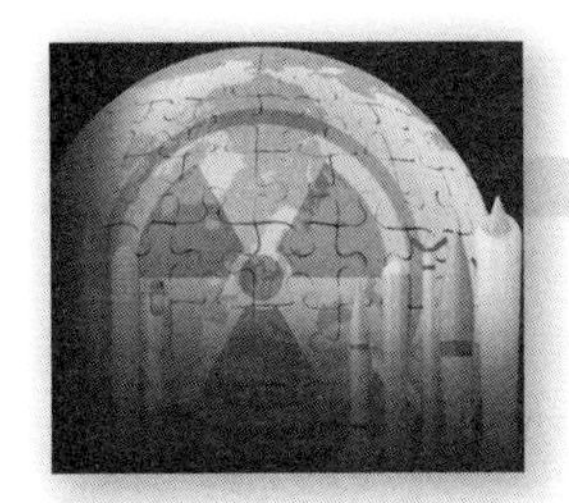

CHAPTER 13

# Emerging Technology and Nuclear Nonproliferation

Today one commonly hears an argument that climate change and nuclear weapons are the two existential threats that our societies face. And while we clearly support the broad awakening of social revelation toward the climate threat, we only wish that even a tithing of that attention would be paid toward its less glamorous, yet probably more terrible, fellow horseman.

The challenges posed by nuclear weapons and the potential for further nuclear proliferation cut across many of the issues we address in our project. Mixing nuclear weapons with the complex landscape of advanced conventional, space or counterspace, and cyber systems explored in earlier sessions raises the possibility of escalation, perhaps by miscalculation, to nuclear use. In our roundtable discussion at Hoover on the changing risks and opportunities of nuclear weapons and nuclear energy, participants lamented the lack of personal memory—in both political leadership and the general public—of the unique danger posed by nuclear weapons. And this was reflected in the turnout to the companion public panel we hosted here for the Stanford University community. Undergraduate students, so eager on earlier questions of social media influence or algorithmic bias, on cybersecurity, immigration, and regional vagaries, were in short supply. This on a campus that fifty years ago hosted an underground network of fifty fallout shelters designed to hold fifty thousand people.[53] As former defense secretary Bill Perry and former Senate Armed Services Committee chair Sam Nunn have both observed, what we have is "the worst of all worlds. . . . We now have all these nuclear weapons, but we don't really have people who take them seriously." Nunn continued, "Complacency is disconnection from reality in the nuclear arena."

We have personal experiences with these weapons. On August 6, 1945, as a US Marine captain on a troop ship in the Pacific, one of your authors and his fellow Marines got word that something called an atomic bomb had been dropped on Hiroshima. By the time they had reached port in San Diego, a second bomb had been dropped—they would not be redeployed for an assault on the Japanese mainland after all. The war was over. And the use of these bombs, along with later atmospheric testing—nearly five hundred tests in the ensuing decades—raised public consciousness everywhere about the terrible power of nuclear weapons.

By the 1980s, President Reagan declared, "A nuclear war cannot be won and must never be fought" and conveyed that sentiment to General Secretary Gorbachev at the 1985 Geneva Summit, along with proposals for reducing strategic forces. And the dramatic meeting in Reykjavik the following year underscored these concerns, with progress toward what became the Intermediate-Range Nuclear Forces (INF) Treaty and the Strategic Arms Reduction Treaty (START). The tides were shifting. Over the next three decades the number of nuclear warheads was reduced to approximately one-fourth of the number in existence at the time of the Reykjavik meeting.

But today, with the Cold War over, the number of weapons held by the United States, Russia, and others remains in the thousands—more than enough to destroy one another several times over—and the circumstances in which they might be used seem to be growing. Equally important, the number of countries with nuclear weapons has grown, most recently to include North Korea, which has tested both nuclear devices and delivery systems. One goal today should be to prevent further proliferation—to Iran, to other countries in the Middle East: no more states with nuclear weapons.

What, our discussants asked, could be done 1) to raise consciousness of the power of nuclear weapons, 2) to develop a more constructive political atmosphere for dealing with them, and 3) to apply new technologies and thinking toward reducing the risk of their use?

## Improving Nuclear Awareness

Through this project, we have identified a number of new challenges that we think deserve public or political attention. But to be frank, the risk of nuclear weapon use cuts across them all. Consider how many news articles you may have read in the past year about nuclear weapons—and how many have you read about Facebook? We have failed to give both our citizens and our leaders a sense of scale and prioritization. The potential damage wrought by nuclear weapons should speak for itself, but Las Vegas casinos no longer market "nuclear" getaway weekends for guests to view the desert tests, atomic high-balls in hand, as they did through the 1950s. And the skies above Los Angeles no longer flash before dawn from nearby nuclear testing. We need to find new reminders that these risks are still with us.

President Reagan's insight that a nuclear war cannot be won and must never be fought should continue to guide us today. How might such a war look?

The effects that would result from military use of nuclear weapons are enormous but difficult to quantify. They would depend on many variables, including the number and size of the weapons used, their technical characteristics, the altitude of detonation, and the nature of the targets. In addition to deaths and injuries resulting from blast and fire in the vicinity of the targets, radiation from fallout would spread over the targeted region, causing further fatalities and severe illness. Widespread destruction of medical, food, energy, and transport infrastructure would lead to many more casualties over an extended period of time. Following a large attack, conditions would get worse before they got better.

Even a "regional" nuclear war would be extremely deadly. Researchers estimate that if India and Pakistan attack each other's urban centers with nuclear weapons, prompt fatalities could reach 50 million to 125 million.[54] Many more would be injured, and casualties would grow still higher over time as a result of destruction of life-sustaining infrastructure and the spread of radioactive fallout.

In addition to this regional catastrophe, recent research indicates a regional nuclear war between India and Pakistan could have significant environmental consequences around the globe. Smoke from burning cities in South Asia would rise to great heights and spread worldwide. It would absorb incoming sunlight, resulting in significantly lower global temperatures and reduced rainfall for an extended period of time, which in turn would reduce food production worldwide.[55] The longer-term environmental and economic consequences, such as food shortages, could ultimately be more severe in terms of death, disease, and injury than the immediate consequences of the nuclear explosions. There are no bystanders likely to remain unaffected by nuclear conflict, even on a regional scale.

A nuclear war between the United States and Russia would have catastrophic consequences for both countries. If their nuclear arsenals were used against both military and economic targets, fatalities are estimated to be from the tens of millions upward.[56] Again, many more would be injured, and casualties would continue to grow over time as a result of destruction of life-supporting infrastructure and spread of fallout.

Here too, in addition to the direct consequences of the nuclear explosions for the United States and Russia, recent research indicates that a nuclear war in which each side employed most of its nuclear arsenal would have a major impact on the global climate, substantially reducing global temperatures and rainfall for several years, drastically reducing worldwide food production.[57] Again, the longer-term environmental consequences of nuclear war between the United States and Russia could be more severe than the immediate effects of the nuclear explosions, and would be felt worldwide.

Beyond the local and global impact of limited use of nuclear weapons is the very real prospect of escalation. The greatest danger of a limited nuclear attack might very well be the high likelihood that other side responds in kind, with no way to know at what point the nuclear exchange might stop. Surely, a nuclear war would be one in which unintended consequences dominate.[58] With vital space-based surveillance, warning, and communications systems likely degraded and information and command-and-control networks likely under cyberattack, misinterpretation and miscalculation could lead to a nuclear war that no one intended. Senator Nunn expressed his view at our roundtable that "we have reached the point where nuclear use by blunder is more likely than nuclear war by premeditated plans."

The numbers are scary, but the point here is not to scare the public or elected officials. Acting from fear is never constructive, in nuclear or other matters. Rather, the goal of reawakening public concern here should be to let facts illustrate the gravity and scale of these impacts in comparison to the short-term distractions that otherwise dominate and manipulate the national political consciousness. And, we hope, informed voters can reward their representatives for prioritizing the existential risk of nuclear weapons in an emerging world.

## Improving Nuclear Governance

A few years ago, roundtable participant and retired bishop of the Episcopal Diocese of California Bill Swing, founder of United Religions Initiative, had this to say about the power afforded to the elected president of the United States:

"You will be like God." That has to sound good to anyone with dirt under their fingernails or the necessity to keep making technical breakthroughs. We all get a little fatigued. How about being a god? That is precisely the option that is laid out in Genesis. "You will be like God."

Or take the phrase "the most powerful person in the world" and ratchet it up a notch. You can put your hand on the Bible at your inauguration, but that is small potatoes. It is when you put your hand on the nuclear trigger and become the single agent of the earth's destruction. That is power beyond human imagining. Take a big mouthful of forbidden fruit. "You will be like God."

Despite the reality of this undeniable power, "gods" is not the first descriptor that springs to mind in contemplating today's denizens of the Potomac. We are humans, and flawed. But humans can build institutions greater than themselves—institutions should be leaned on in an effort to improve the governance problems of nuclear risk.

Our roundtable contributors described the political space today in Washington for constructive interactions with Russia as poisoned by the domestic fallout from Russian influence efforts in the 2016 election. Under the Trump administration, the executive branch, which would normally lead here—consider Ronald Reagan and his cabinet's personal contributions to the arc of Soviet negotiations in the 1980s—is not seen as credible on the topic. That leaves Congress.

Contributors argued that Congress has to step up on this problem by interfacing with relevant executive branch agencies through regularly scheduled security policy conversations. A full discussion of reforming the dysfunctional War Powers Act is beyond our scope here, but in general we think that it would be a positive step for Congress to reassert itself into national-security and foreign-policy decision making and to build the institutional infrastructure and processes to do so on a regular, meaningful basis. As Sam Nunn summed it up in our roundtable discussion, "Congress under the Constitution is supposed to make the decisions on war and peace, and yet in the one type of war that could end the world as we know it, Congress has virtually no role. But an irresponsible role is not going to work either."

One concrete step would involve the creation of a bipartisan "liaison group" from the House and Senate, with members appointed by the leadership rather than committee chairs (given the reality of how weak chairs have become). To permit the degrees of freedom that would be needed for substantive action, the liaison group's remit should be relatively broad: include not just nuclear arms control, but NATO and the US-Russian relationship in general. Geo-economic considerations would play here, too—contributors observed, for example, how the matter of Ukrainian national security is intimately tied to its economic security. This is exactly the sort of group that would not be able to issue press releases touting one's win or another's loss—but it would improve the flow of information and help establish an understanding. And it should actually be welcomed by an administration looking to get votes for eventual legislation enabling its policy goals or international agreements. Recall again President Reagan's words of advice toward this sort of consultative process: "If you want me on the landing, include me on the take-off." Developing a better understanding of views toward extension of the New START treaty, which is otherwise set to expire in early 2021, would be a good first assignment.

Reestablishing military-to-military communications on nuclear and other threats was also identified as a major US-Russian opportunity in the current environment. These avenues for discussion, and mutual airing of concerns, have contributed to stability over decades, even as national political relations have waxed and waned. Congress has largely cut off military contacts in recent years as punishment for Russian adventurism, but uniformed military leaders in both countries nonetheless report continued interest in improving mutual communications. Military-to-military connections help reduce the risk of an inadvertent crisis, and they also help build the groundwork for broader agreements over time. In our recollection, for example, the ability of President Reagan's US State Department and Soviet foreign ministry negotiators to make substantive decisions in any security policy meeting was always predicated on the participation of Soviet military representatives. Reykjavik was one such example of having all the right people in the room. And while the United States' military-level communications channels with China today are much stronger than those with Russia, our project discussants have pointed out that even US-Chinese military communications are substantially below where US-Soviet links were during the height of the Cold War. Congress should give military commanders the encouragement to engage in any such discussions that our uniformed officers deem valuable.

This issue of expanding the nuclear discussion beyond the United States and Russia was another topic of roundtable discussion—and a source of audience questions at our public panel. On the one hand, the United States and Russia still have 90 percent of the world's nuclear weapons. So what we do matters. On the other hand, given broader military capabilities across many actors and the emergence around the globe of nonnuclear disruptive technologies that implicate nuclear weapons, what other countries do also matters, beginning with China. Policy minds in the United States and Russia pointed to Chinese missile technologies and deployments as a reason to leave the INF Treaty, unconstrained as China was by the treaty.

Even changing demographics come into play. At the height of the Cold War, the combined number of people in the United States and Western and Eastern Europe (including the USSR) held at risk to the nuclear threat was about seven hundred million. Today, in South Asia alone, where Indian and Pakistani arsenals continue to increase in size each year, that number is 1.6 billion, and rising past two billion by midcentury. Nuclear weapons are now very much a global concern.

As roundtable contributor and Carnegie Endowment for International Peace analyst Ashley J. Tellis described in his remarks on the subcontinent, the dynamics of and the derisking strategy for every regional nuclear dance cannot be put in US-Soviet-deterrence terms. India views its nuclear pursuits differently, and as arising out of a different historical context, than Pakistan does. India compares its own nuclear stockpile as an insurance policy against the capabilities of its neighbor China (which in turn calibrates its ambitions vis-à-vis survivability against US and Soviet capabilities). Pakistan, meanwhile, positions nuclear weapons as a potential complement to its outmatched conventional forces in an existential regional conflict with India—some voices within the military-dominated country further view the capability as a sort of umbrella that licenses it to conduct its own cross-border terrorist activities. Misinformation and secrecy shroud every step for these states, who both lack the international legal recognition to conduct an open nuclear weapons program. And given their relatively late starts, neither party views itself as "sufficiently nuclear," which points toward an expected direction in nuclear armament—that is, toward more—if not a desired end state.

Given this idiosyncratic landscape, and the fact that we are only secondary actors in this drama—our interests therefore often viewed with skepticism—the United States has so far had limited progress in South Asian nuclear security. The United States' efforts with Pakistan after 9/11 to secure its nuclear facilities and pursue al-Qaeda were crucial (and welcomed, given that country's ongoing internal security concerns), but financial and military support has not extended to an enduring broader relationship. And India has not responded with enthusiasm to direct offers from the United States to help improve the security and safety of its weapons-handling programs—Indian officials have at times interpreted US bilateral advising as unwanted lecturing—but it has shown interest in learning from international best practices when presented in a neutral, global context.

The US-Indian civilian nuclear power agreement negotiated by the George W. Bush administration, while rocky in implementation, was one good example of positive creative diplomacy. It was a practical effort to work out some form of stability and a platform for future bilateral cooperation and investment, despite India not having followed formal international routes toward its nuclear program. The United States should continue to explore ways to make our participation regarded as valuable in the region. We could, for example, convene international summits focused on the potential effects of emerging technologies on nuclear security—cyber threats to international nuclear monitoring and detection systems, for example—with South Asian participation.

Another enduring opportunity is of course economics. Roundtable discussants remembered leading so-called track II security dialogues of business, civil society, and former military and political leaders in the region, during which both India and Pakistan recognized the immense mutual gains that might result from normalization of trade and commerce across their border. And there have been attempts over the years to this end, scuttled in part due to the undulations of regional politics, and perhaps more fundamentally by Pakistan's fears that its economy would be overwhelmed by Indian competitors. So it's not an easy problem. But given the overall gains that could be achieved—to say nothing of follow-on regional stability dividends—economic normalization is a question of when, not if. The United States should have this on its agenda as it looks to navigate dramatic changes in global trade attitudes (and technological options) across an emerging world.

Finally, the United States can aim even higher: it could use what goodwill it has in the region to consistently encourage an overall Indian-Pakistani peace process. Roundtable discussants with long experience in the subcontinent were grounded, yet hopeful for progress. This initiative would require the involvement of not just Pakistani civilian leaders, who have at times been open to rapprochement, but its military, too. And the Indian state is understandably skeptical of dealing with the Pakistani state and military. We cannot compel either side to sit down and have an agreement—but at the same time the United States has never shown the interest or wanted to invest the political capital in getting both sides to think about a solution. That may change as the United States electorate itself does: as of 2017, India ranked second to Mexico as the largest source of immigrants living in this country, second in naturalizations, and first in new arrivals. The only plausible long-term solution is for both sides to learn to live with what they have already got—and in the meantime, we should be looking for ways to help "keep the pot from boiling over."

More broadly, uncovering the best approaches to nuclear governance in the world beyond the United States and Russia will in each case require experimentation. Creativity in our own international arms-control efforts should be encouraged. The Trump administration's initial efforts to break through the North Korean situation via direct high-level consultations, for example, should be applauded for moving the needle. But without follow-up, that progress could disappear. Similarly, the state-by-state, situation-by-situation efforts to secure nuclear materials from around the world over the past decades have gradually reduced the number of countries in possession of such materials from forty-two in 2001 to twenty-two now. Russia was a very cooperative partner in that. That was not a flashy breakthrough, but it was good management, and it has reduced risk.

## Improving Nuclear Technologies

Our contributors described how new weapons technologies—hypersonic missiles, cyber intrusion, or the use of AI in decision making—could be destabilizing forces in the nuclear landscape. But could new technologies help reduce new risks in this emerging world as well? If they could, then the right combination of mitigation and ingenuity could improve the situation.

Consider that interest in nuclear technologies broadly is likely to grow around the emerging world. For example, civilian nuclear power represents just 10.5 percent of global electricity production today, and less than 5 percent of total energy consumption. And while costs remain a barrier today to rapid adoption, new packaging technologies—such as small modular reactors, or the new chemistries of so-called Generation IV reactors—could surprise with their appeal in a carbon dioxide emission–constrained world. Countries who already use nuclear power today use it to meet, on average, less than one-seventh of their electricity needs; expanding that use would likewise increase the volume of nuclear-fuel-cycle materials that will need managing. And countries representing one-quarter of global electricity demand currently have no civilian nuclear power, despite their having the long-standing international right to develop or acquire such technologies under the Nuclear Non-Proliferation Treaty of 1968. This includes, for example, most Middle Eastern and African countries where populations and energy demands are expected to grow rapidly, but which lack substantial experience with the technical or institutional management of nuclear fuel cycles. Iran, for example, built the Middle East's first nuclear power reactor in 2011, and it was quickly followed by the United Arab Emirates and Turkey, each constructing four units of their own, all using conventional nuclear technologies. Iran is now in the process of developing two additional reactors, as are both Saudi Arabia and Jordan. In just over a decade, the region will go from no nuclear power states to five.

Former secretary of energy Ernest Moniz in his paper prepared for our discussions enumerated four areas of opportunity: detection, prevention, the rollback of proliferative behavior, and responding to subnational risks. Moniz described, for example, how machine learning, applied across syncretic platforms such as Palantir's that specialize in finding patterns across disparate data sources, could be used to identify proliferative behavior in supply chains and global trade patterns, automatically scanning across Twitter posts, social media images, cargo ship manifests, money transfers, satellite images, and more to infer underlying weapons-related activities. This ability becomes more credible as more aspects of society and economic interactions are digitized, and therefore accessible to computer interrogation.

Moniz pointed out that leading-edge technologies—in satellites, sensors, and networked systems—have long been relied upon to verify state behaviors around the treatment of nuclear materials, and we should continue to push that envelope. Improved forms of detection and verification could give us the confidence, for example, to partner in civilian nuclear power deployments in the Middle East without a legally binding obligation not to enrich or reprocess.

New technologies and approaches could also help work through existing barriers in nuclear waste management. The United States faces its own political stalemate over the disposal of civilian reactor wastes, as well as the transport and storage of higher-grade military wastes. The Yucca Mountain long-term disposal site, for example, has continually been described as a political problem, not a technical problem. Nonetheless, politics have failed to map a path forward. Technology might. Moniz's suggestion of pursuing deep-borehole disposal for military wastes, combined with an interim (e.g., one-hundred-year) dry cask storage system for civilian waste, would effectively buy more time for the federal government to improve safety, help reverse today's negative public attitudes toward nuclear power, and open up the political space for further deliberation.

Leaning more heavily on technology to reduce the risks of nuclear weapons and materials will of course require technologists. In the civilian sector, the demography of nuclear workforces is already a concern, with power engineers in the United States aging alongside their 1970s-built reactors; start-up nuclear energy programs around the world are on the hunt for experienced talent from the United States and Europe. Senator Nunn used our public panel to implore today's university students, who so often say they want to use their technical skills in more meaningful forms of work, to consider how their careers could contribute to reducing nuclear risk. "There are ways to do this!" he urged. "There are ways to make the world safer." Nuclear security can seem like a distant topic, the sole province of national governments with opaque alphabet agencies, but an increasing role for technology in this space should open that door to nonspecialist contributors and private-sector digital technology expertise. Nunn continued, "And if it were up to me, I would challenge Russians to do the same things and then see where we can meet—maybe even find ways where their technologies or our technologies can work together." He's right, and it's an idea worth trying.

Concluding on the topic of risk, Nunn quoted Warren Buffett's back-of-the-napkin math on the nuclear situation. "If an event has a 10 percent chance [of occurring] in a given year, and that chance persists for fifty years, then there is a 99.5 percent chance that it will happen over that fifty-year period," he said. "But if you can reduce that 10 percent chance to a 1 percent chance, and that chance again persists for fifty years, then you have a 67.5 percent chance of it not happening." The point is that nuclear weapons, as long as they exist—as with other emerging existential risks—are not solved problems. They are, as we used to say in the Bechtel construction project management days, "work-at problems."

What do we mean by that? Say for example that you are given the job of building a bridge across the Potomac River. So you lay the foundation piles, pave a road across—job done. But now, what if the task isn't just to build that bridge but instead to do so in a way that also minimizes any worker lost-time accidents in the process? Now, the job is no longer a simple solvable problem—it's a work-at problem. Putting up a safety sign at the site gate isn't enough. Instead, you have to keep your objective in mind with every small step you take. So, too, nuclear weapons risk reduction might not always look like a sweeping resolution as much as a continuous series of step-by-step improvements. Compounded over time, this makes a difference. In the nuclear world, we should be seeking out wins, and holding on to them.

In this project, we've invoked the giants who led the United States through an earlier hinge of history to develop a constructive framework for dealing with the world's—and by extension, America's—problems in the days following the destruction of the Second World War. Given the new global changes we can now observe, if we were to go back to those Truman, Acheson, and Eisenhower days, what would they be thinking? What would they do?

President Eisenhower, in his 1953 Atoms for Peace speech, laid out a strategy of nuclear management that recognized the technology's great social potential while also aiming to limit its risks. That "weeding" process, which authors of this volume have played a part in, has continued to this day. We need to continue to incorporate the latest technologies and thinkers in service of reducing those risks, in creative consultation with our long-standing global allies and with our adversaries—so that the next time the American people forget about the nuclear risk, they can be justified in doing so.

For more:

- "Nuclear Nonproliferation: Steps for the Twenty-First Century" by Ernest J. Moniz
- "Nuclear Dangers in an Emerging World" by Sam Nunn
- "A Troubled Transition: Emerging Nuclear Forces in India and Pakistan" by Ashley J. Tellis

*https://www.hoover.org/publications/governance-emerging-new-world/fall-series-issue-919*

**From "Nuclear Nonproliferation: Steps for the Twenty-First Century"**
*by Ernest J. Moniz*

Advances in big data have other significant implications for the nuclear detection mission. In the longer term, the vast quantities of publicly available data might be used to supplement traditional means of verifying international agreements. Traditionally, proliferation detection has been the exclusive task of governments and international organizations like the International Atomic Energy Agency and the Comprehensive Test Ban Treaty Organization. Monitoring and verification of proliferation and arms-control agreements have historically relied on tools such as monitored party declarations, on-site inspections, and national technical means. . . .

Last year NTI [Nuclear Threat Initiative], in partnership with the Washington-based nonprofit group C4ADS, initiated a project to determine what might be possible in terms of using publicly available data for proliferation detection and monitoring. This project aims to demonstrate how use of publicly available data and network-analysis techniques can supplement traditional monitoring and verification of international agreements. To do this, the project takes advantage of the large quantities of data that are generated around the world every day (e.g., financial records, scientific publications, property records and registrations, import/export data, shipping data, etc.) as well as novel, proprietary technologies (e.g., Palantir, Windward) that permit network analysis and facilitate investigation of possible illicit activity.

The initial phase of the project is nearly complete, and the early results are promising. The data analysis has allowed the project team to detect trade in dual-use items. It has also been possible to determine whether freight forwarders are used and the location of an entity within a trade network. More broadly, mining these data permits researchers to generate trade profiles for companies, setting comparable baselines to help determine what sorts of trade and activities appear "normal" and which ones indicate illicit or proscribed actions.

The overall goal is to characterize "risky" trading patterns that could then prompt a more detailed analysis.

PART

# V

# What Have We Learned?

We now have fully explored the impacts of the hinge of history on the United States and on other countries around the world and have examined across-the-board issues as well. Clearly the impacts are large, and equally clearly we have an opportunity, with a major effort, to take full advantage of these developments and mitigate the problems—while neglect knowingly or unknowingly will likely lead to many disasters. Most of how we navigate the emerging new world will come down, first, to individual people solving for the problems and opportunities they face each day. And, second, we will need better domestic governance. But third, somehow, we also need to regain a sense of the world community—and the way how, block by block, there can be built again a new security and economic commons.

We see that new technology has tremendous promise to increase productivity and enhance our way of life. We also see a very widespread pattern of disruption. We see major demographic shifts, namely among the young and growing developing nations and the aging developed states. But there is interplay between changing demographics and advancing technology. Each offers substantial benefits and poses challenges, and in many cases, one offers potential solutions to problems raised by the other.

We see the disruption of the "Bangladeshes" of this world because their low-cost labor is threatened by automation. We see a deglobalization taking place. We see the scary prospect of inexpensive, lethal weaponry, which makes any fixed installation a target open to attack and which we will need to counter. We see the need for a lot of retraining of people to go from job A to job B, and improving the quality of K–12 education. We see the spread of social media and other communications technologies, which complicate the challenge of governing over diversity. We see the restructuring of global populations, with potential for prosperity or disaster.

Importantly, we see how these changes may create great opportunity for the United States, compared with other parts of the world. How do you galvanize the country and the world to recognize these opportunities and problems and to do something about them? We think that it has to start at home. Leadership from the United States will be essential, and it seems as though the United States will be the one in the strongest position to do this. But delivering that will require an emerging conviction from within that this problem is one that will require the United States to be engaged in the world. Institutions that were built up after the Second World War—NATO, the World Trade Organization (WTO), the International Monetary Fund, and so on—even if in doubt today, have nonetheless already given us the habit of how to work together. And we would be making a huge mistake to not recognize this and in so doing turn our back on today's constructive options.

How does the United States position itself to take leadership in maximizing the advantages of artificial intelligence, additive manufacturing, and other new advanced technologies? We could focus on competition with China. But that's the wrong way to look at it. The more ideas that are developed anywhere in the world, the better for everybody, assuming we can have some degree of reasonable transparency. When the United States launched the Marshall Plan after the Second World War it was in a sense looking around to a weakened world, with uncertain influences. The pitch then wasn't one of outcompeting the other countries per se. Rather, our challenge to those who we knew would be part of the fabric of this emerging new world was: turn your country around so that you can be part of the solution, not part of the problem.

Under these circumstances, then, and with this hinge in mind, what should we do?

First of all, we need to figure out how to make these issues visible and talked about on a global scale. This is a new situation we wish to convey—but, in a way, many of its characteristics are familiar.

In the fall of 1985, one of your authors was on his way to Moscow to meet with Foreign Minister Eduard Shevardnadze and Mikhail Gorbachev in preparation for an upcoming summit with President Reagan in Geneva. Gorbachev didn't want the meeting to just be about various conflicts between our two countries, but rather to set the stage for something bigger, and still important to the both of us. How should we be engaging with this other superpower on matters of substance? Shevardnadze earlier had conveyed the sentiment that whatever the Soviets did—on human rights, arms, reforms, and so forth—would be done in their own interests, not out of some obligation to US entreaties. And he was right to say so. At the same time, while your author had gotten some personal sense of the new general secretary, our intelligence agencies struggled to understand how Gorbachev saw the world and the Soviet Union's place in it. We were particularly interested in how he thought of its closed and compartmentalized society. Was Gorbachev satisfied with that system's undergirding future Soviet growth prospects?

It was clear even then that the information-age technologies emerging at the time would fundamentally alter the global landscape in terms of freedom of commerce and expression, and in driving a new wave of innovation and creativity. Given that, it would be in the best interest of the Soviets themselves to adapt to these changes, moving toward a more open society that could harness these forces, rather than falling further behind the West in attempts to suppress it. So there were clear implications for governance. Your author described these changes in some detail—some of our own advisors unfavorably termed it "a classroom in the Kremlin"—and asked of Gorbachev, "What sort of world do we want to build for the future?"

Somewhat surprisingly, Gorbachev engaged on the issue. He even joked about joining the Soviet economic planning office to give them some new ideas. It was clear to us by that point that Gorbachev was not someone who was looking to overturn Communist systems. In fact, he would give sharp defenses of its products, and of the health and capabilities of his economy. But he was wrestling with the broader ideas of governing over a changing world. Our US diplomatic team would try the same argument again a couple years later in a separate meeting with Shevardnadze, who took notes that he later confided had been relayed to the Politburo members. And over time this way of thinking took hold. By the time of Gorbachev's 1988 speech to the United Nations General Assembly in New York, he was speaking of a "radically different" world, where Soviet society would have to become "Democratized in practice . . . [with] perestroika now spread[ing] to politics, the economy, intellectual life and ideology."

This is the power of such ideas. And in this new hinge of history, with even more profound changes to the governing environment, it's important that we continue to spread them. We believe the best candidate for this role is the United States—picture the president of the United States, flanked by key members of Congress, perhaps back before a special session of the United Nations. We should outline for the world our own principles, and follow up by then helping each country develop its own way of response, too.

What should they say?

Through more than a dozen sessions aimed at understanding the impacts of changing technologies and changing demography in the emerging new world, both around the globe and at home in the United States, a few themes emerge around what should be done, and what should not be done, in response.

## We Need Many of the Changes That the Emerging World Will Bring

As wryly observed by one of our repeat project discussants, the list of crises one can worry about can only get so long. It's easy to look at the world as it exists today, focus on some change that might affect a particular place or set of people, and start to worry. The secular changes we see coming in this emerging world by nature bring us into uncharted territories with many challenges.

But the good news is that when people, businesses, and institutions are allowed to adapt, they cannot help but do so. The world is not zero-sum. Certainly growth and experimentation and change are the desired default around much of the world, which is developing from an unacceptably poor current reality. And our economy and society in the United States is not zero-sum, either.

Advancing technologies provide for new productivity, economic growth, and a higher standard of living. New technologies can be disruptive, but we have to resist the urge to restrict them and find ways to capture their benefits and mitigate the problems. In fact we are relying on them to extract us from some of today's deep structural problems, which may otherwise consume us.

Consider the defined-benefit entitlement obligations, such as pensions or health care for the elderly, which were freely granted in an earlier time but whose bills are now coming due amid an aging population. If we continue business as usual from the comfortable present, trillions of dollars in unfunded liabilities are set to bankrupt cities and states across the country, or at the very least squeeze out any budgetary room for elective public services. The US federal government, meanwhile, is dead broke. Its discretionary share of spending decline has already declined from 62 percent to just

31 percent over the last fifty years and is set to shrink further as mandatory spending explodes. Not fixing this would be an intergenerational disaster, in plain sight and of our own doing, that would make laughable the amount of time we spend today concerned with nefarious bots on social media or the safety of driverless cars. But if we are able to take advantage of the seemingly magical economic dynamism and productivity on offer through emerging technologies—together with necessary but modest entitlement reforms—then this becomes a problem we just might manage. China stunned the world in pulling hundreds of millions of its people out of seemingly intractable poverty through a once-in-a-century transfer of rural farmers to urban manufacturing centers. Could today's scalable technologies give Americans a similar productivity boost? Growth of that sort could provide the United States government of the future with the fiscal capacity to do more than mail Social Security checks and file Medicare claims. We should grasp this opportunity, as should others around the world.

A similar argument could be made for the benefits on offer to low-fertility, rich countries, who otherwise face drastic workforce and population contractions, from a global pool of migrants willing to shore up their citizens' ranks. Here again, to continue with the status quo toward a Malthusian daydream would be the radical choice, certain to disrupt today's way of life. We have experience with population declines in human history—the aftermath of the Black Death in medieval Europe, the decrease of Salvadorian men following that country's civil war, or, to take a more limited example, the tribulations of inner cities or small towns of America with boarded storefronts. None is worth repeating. Through this project, our discussants have described a country's demographic prospects as utterly knowable, certainly in comparison to the vagaries of economics or future technologies. But the United States is fortunate to in fact have an option to control its own demographic destiny across the hinge of history by deliberating a strategic immigration policy that complements its organic population trends to sustain our economic growth.

At the heart of our focus on governance is an idea that good policy aimed at the changes of the emerging world can help us take advantage of their opportunities, while mitigating their problems. But this doesn't mean that government will—or should—control how that unfolds. New ways of communicating, new ways of producing goods and services, even the movement of peoples cannot be regulated into existence by fiat. Ultimately the benefits of these innovations are brought to the people by the market. And the market works very well. There is always a temptation to intervene, and obviously some laws and regulations are needed, but there is a balance to be struck between attempting to guide these forces and snuffing out their spark. It's always easier to focus on the imagined bad, happily avoided, having missed out on the unknown good, regrettably prevented.

## What Do We Owe Individuals, So That They Can Benefit from These Changes?

Part of today's social angst stems from a feeling among many Americans and others around the world that they are not sufficiently equipped to confidently deal with the emerging-new-world problems that are coming at them. And they are right. The most important thing that governments can do in response is to provide better K–12 education that gives students the capability for lifelong learning and self-determination. This is true in Nigeria, where data shows that girls who do not receive at least a middle school education are not able to effectively plan their own family structures (and the ensuing allocation of family resources and time). It's also true in Germany, where our contributors have noted that success in later-life retraining actually depends upon an individual's basic schooling; responding to public shock over international testing results that placed German schoolchildren in the bottom 10 percent of students among OECD member countries, the country has in less than two decades climbed to the top quartile. And it is true in the United States, which suffers from continued poor performance overall, where policy and resources too often seem focused on the interests of the adults rather than the students in a school, and where many American parents have no choice but to send their children to low-performing or unaccountable schools.

Outcomes here have huge human dimensions. A historical analogue: when the impacts of the European Industrial Revolution swept through a long-inward-looking East Asia in the nineteenth century, many countries were caught unprepared and were unable to respond. Many were even colonized. One exception was Japan, which with its Meiji Restoration undertook major proactive social and economic reforms and brought in foreign advisers to train its people in these new technologies. Less appreciated is the role education played. How was the completely closed Japanese society able to make such quick use of these alien technologies and practices, taught in foreign languages by strange outsiders, such that they would even be able to fight and win wars with advanced Western powers by the end of the century? Consider that in the early 1800s, the literacy rate in London was less than 50 percent. In Paris it was under 30 percent. But in Edo-era Tokyo? It was an astonishing 70 percent.[59] Japan's widespread basic education levels gave its common people—and not just the small share of elites who could attend newly established engineering colleges—the foundation to succeed and learn new skills in a rapidly changing world.

Following K–12 schooling, many continuing students in today's United States are fortunate to have access to what is probably the world's best university system. Even here, however, students should be exposed to better information about the career prospects and associated earning potential of the majors they choose compared to the lifetime costs they will incur from borrowing to pay for that degree. Having up-to-date data will become more important as the technologies of the emerging world increase churn in the workforce.

How do we reach back to those people in our population who need a second shot because they weren't served right by our education systems in the first place? For one, students who do not enroll in four-year universities should be presented with a choice of vocational training opportunities or access to efficiently run two-year community colleges that offer career technical training that leads directly to skilled employment, or two years of low-cost, high-quality college education. The norms of vocational training vary around the world and by industry, with technical apprenticeships used to good effect in Europe and in Japan, for example. Again, the United States can make good use of its existing community college networks in the emerging new world by focusing on their value for students: maintaining low-cost structures through efficiency in administration and a focus on teaching skills, cooperating with the local private sector to develop up-to-date curricula for more employable graduates, and developing flexible delivery, such as online courses that encourage enrollment by students in different stages of life and jobs. And community colleges foster the habit of lifelong learning that will be essential for success in tomorrow's economy.

American baby boomers on average experienced six spells of unemployment over their prime working years. And each such instance comes with a reduction in lifetime earnings and a chance of falling into poverty. If we believe that the technologies of an emerging world such as machine learning, additive manufacturing, or robotics will lead to even more rapid changes in the workforce, then it is also important that our social safety nets become more focused on getting those who are laid off back into productive work, quickly. Job-loss programs should encourage rapid reemployment, retraining between jobs, and the domestic movement of willing workers to willing employers throughout the country. And governments should reduce barriers to switching careers, such as onerous occupational licensing or other regulatory barriers to internal migration.

And while we are focused on what governments can do to equip their citizens with the faculties to take advantage of the emerging new world, it's important to recognize that firms also share some responsibility as the front-line mediators among technology uptake, productivity, and their employees. Firms also have better information about what sort of disruptions they do or do not face, what skills they do or do not need from their workers, and where new opportunities lie. And they move faster than governments do. Governments should view them as partners and complements in education, in training and retraining, and in providing data on workforce needs.

## Collecting Data Can Help Us Manage the Complexity of the Emerging World

Twenty-first-century technologies—namely cheap sensors, pervasive mobile computing, scalable cloud storage, and machine learning—make the collection and use of very large data sets cheaper and easier. Usable data drives the notion that you don't have to rely on past practices or theory alone to make important decisions. And time and again, our project's discussants have illustrated the way that data use creates a positive feedback loop to take advantage of the changing world. To date, much of the focus of data collection has been on personal consumer data for digital advertising, and adjacent privacy concerns. This is something that people can see in their day-to-day lives, and there is a clear market value to it, so the practice has become very salient. But the universe of important problems that can be addressed with richer data, on both people and their surrounding environments, goes far beyond consumer preferences.

For example, we've mentioned the plight of the world's "Bangladeshes," which are susceptible to the potential loss of low-cost labor advantages that have led to the development of those countries' industries. But the fact is that despite Bangladesh's having faced all manner of challenges—ethnic diversity, floods, cyclones, famines, climate change, shifting supply chains—it is actually dealing with these problems with remarkably positive results, and playing a poor hand well. Bangladesh's weak if essentially benevolent government has presided over decades of surprising improvements in family planning, in health care, in education, and in economic growth. (It was certainly surprising to talk with experts on Africa, for example, and to hear their admiration for Bangladesh's ability to build institutions under duress.) It has done this through good governance that took advantage of what was offered to it to by nongovernmental donors—and through open and reliable data. Bangladesh has essentially outsourced many public services to the civil sector, domestic and international, but its government has maintained an extensive public statistics capability that gives transparency on the results of these efforts—what works and what doesn't. Over time this improves accountability, and it encourages further outside philanthropy and investment.

Or consider American K–12 education, where data on school, teacher, and pupil performance can be used to improve transparency. In Part III of this book's examination of the US economy, we saw how high-performing schools collect extensive data on student performance, which is shared with teachers and used for continuing improvement. Low-performing schools resist collecting data on student performance, for fear it will be used against teachers. Broadly speaking, better information improves the functionality of the marketplace for education populated by parents and administrators: we know there are good schools and there are bad schools, and transparent data sharing about this sets the starting point for reforms. It can also help to personalize education—identifying weaknesses at the individual student level for quick remediation before falling further behind, or promoting those areas where a student is excelling and is ready to move ahead. And analyzing student performance data across classrooms and even schools, for example among niche student demographic populations who might otherwise be seen as outliers within a single classroom, offers a promising path forward in identifying the most effective teaching and school administrative methods, so that they can be scaled beyond one-off "Camelot" schools.

We've also heard how community colleges can use data about their graduates' performance in the labor force to let students make better decisions about what to study. That return on investment is particularly important for nontraditional students who may be taking time off from work with the express purpose of upgrading their skills.

In national security, experts and technologists have described how the side with the best and most reliable information will have an advantage in the digitally enhanced conflicts of the emerging world—and how that is already driving competitive global investments in civilian information infrastructure and standards today. It's important that the United States military develop the flexibility needed to recruit the country's best and brightest minds to contribute to that expanded responsibility. And it is important to reduce the barriers to entry for smaller technology firms that seek to contract with the Pentagon, encouraging more rapid acquisition of commercially available technology and reducing onerous contracting restrictions.

Finally, we wonder if data may help reverse what we see as negative trends in the structure of American democracy. Our project's contributors have lamented how, in the United States, both the scope and complexity of governance tasks that motivated citizens and organized interest groups have demanded since the 1970s have led to a transfer of political power. Power has shifted away from cities and states and toward the federal government in Washington, DC. And within the federal government itself, this shift has led to an enlargement of the executive branch's declarative professional regulatory agencies at the expense of traditional congressional deliberative processes. The result is stagnation on key policy questions, where all politics become national, all policies become partisan, and even where there is room for compromise—and there usually is—the opportunity is allowed to pass.

Against this, data offers hope in two ways. First is the possibility for Congress to make better use of information and analysis through data. In recent times, the agencies have taken over so many congressional duties in part because they have the staff, budgets, and expertise to handle data analysis that is as complex as the emerging US economy is. As data becomes cheaper and easier to use, Congress should invest the resources needed to gain the confidence to make more-detailed determinations on its own, accountable and in the public eye, rather than delegating major "technical" implementation decisions to the legalistic and often stakeholder-dominated administrators and bureaucracies.

Second is the potential for a data-enabled "new federalism." In a speech for President Nixon's administration in 1970, one of your authors used this term to describe how even then the country had become too centralized in its approach to governing, and how we as citizens had in the process absolved ourselves of our local or even personal responsibility to solving complex problems. Since then, much of the governing responsibility that has continued to be transferred to the federal government has been done with the argument that the complexity of problems demands the sort of resource-intensive and highly skilled decision making that local leaders in cities or counties or even states lack. But data and machine-learning analytical tools could help localities take on more governance questions in-house by becoming good enough (to borrow an earlier phrase from our discussion of African governance) on the tricky analytical questions while adding in the—often more important—attributes of timeliness and local suitability. The federal government should explore ways to support local authorities in improving their technical decision-making capabilities, and in sharing its own data and tools and resources with them, to off-load responsibilities where appropriate. States and cities can be laboratories for experimentation and innovation and development of practices that work and can be shared. From our window in the early part of the 2020 COVID-19 pandemic, we see that state or even county leaders around the country are being prompted to step up and make complex, locally informed, data-informed decisions with no right answers, and with life-and-death consequences—from both public-health and economic perspectives. It seems that this process is well underway. And they are learning from one another.

## The Personal Responsibilities of Digital Citizenship in a Networked Age

About 80 percent of American adults are now thought to use social media, and we see this as an essentially positive force. We believe that new technologies have given Americans ever more abundant choices in terms of how to spend their time, how to communicate with one another, and how to learn about their world. So the fact that they are choosing to use these network platforms for an increasing number of hours each day suggests that they find value in doing so. This is a not a problem for governments to solve.

What is different about network platforms from previous technological changes is that they make all people not just consumers but also publishers of information. Not just buyers of content but sellers of their own attention and personal information in exchange for the services provided to them. This gives ordinary people new responsibilities that they may not be accustomed to, where even low-effort actions can affect their own well-being and that of others. Using social media is not like watching TV. For example, platforms make it relatively easy to willingly or unwillingly spread disinformation across large audiences, which is particularly problematic in free societies that rely on good information for the function of free commerce and free elections.

Responsibility and accountability for one's actions are core American values that can shed light here. And existing US law put in place from the birth of the internet essentially enshrined the idea that the operators of digital platforms and network services need not be the gatekeepers for the actions of their users. But the implicit digital markets in which such exchanges occur—of content, of opinion, of political beliefs, or of personal data—remain poorly defined. Most individual participants likely do not even understand that they are in a market at all, let alone that they face some duties as part of that. So what we are left with is a situation in which essentially no one is accountable.

Some have argued that the response to this should be to reimpose responsibility on the large developers and operators of these platforms. But this is an incomplete remedy, and a parochial one. Instead, why not find ways to better inform users—and even compel them to learn—about the impacts of their social media choices, something that is essentially opaque to them today? Platform operators should share some duty to ensure that they are transparent about the sorts of transactions they are implicitly involving users with—the sharing of personal data, for example, or the amplification of their speech. And platforms should offer their users credible choice as to the extent of that transaction. If a user wishes to share less personal data while using a digital service, for example, then they could be offered an equivalent-value financial option, or they could see a degraded experience. To police this, the government, meanwhile, could certify that barriers to entry in these very-fast-moving digital markets remain low enough that consumers do have real choices among network-service providers, for example through personal-data or social-graph (i.e., your friend/follower network) portability requirements.

Users, meanwhile, should also see accountability for their network platform actions. Here, the solution is less generalizable, but you could imagine how private platform operators might enforce positive and negative social behaviors through nudging, directed feedback, social pressure, revocation of privileges, or other behavior-targeting design solutions, many of which already form the backbone of digital service user-experience design. From a regulatory standpoint, a platform operator's ability to transparently demonstrate that it has made good-faith attempts to manage its user base could be a requirement of continued shielding from liability for user actions. Individual judgment questions of policy (Should social media allow political ads? Should platform operators fact-check speech?) would be decided by the provider. Of course, here again, it would be imperative to ensure that users are exposed to a competitive market for such services, so that providers could be rewarded or punished based upon user satisfaction with such self-regulatory regimes.

Social media is not an inherently flawed product, and we think that there is more risk in too much attention being paid to "fix" it than in simply trusting in the American people to learn about these new tools, and over time take it upon themselves to work out the kinks. American design and ingenuity created these tools, and that market moves very fast—far faster than governments generally do. So long as competition persists in the sector, new ideas are likely to be rewarded if they further refine how these services function, through entry and exit, or by pushing incumbents to up their own games through self-disruption. We have not seen the end of new communications tools, or the final forms of social media and network platforms. This is an open field.

## Combining Technology with Demography as Complementary Strategies

Our project has considered the new challenges posed by changing demography around the world. And we have discussed the impacts on people's livelihoods from emerging technologies. And, the major governance challenges that arise from these shifts, in a time when many governments are weak. One thing that makes us optimistic about the path forward, however, is that some of the problems of demography can be helped by technology, and, in turn, that the problems of technology could be ameliorated by demographics. In a sense, they can be complements.

Consider the growing use of additive manufacturing and other advanced manufacturing techniques. These technologies will make it easier to create things near where they will be utilized, and to provide flexibility in global trade patterns by introducing new degrees of freedom as desired by the manufacturer—in particular, reducing the central need to optimize labor costs. Depending on policy priorities and where these capabilities are used, this could lead to a domestic reshoring of manufacturing capabilities, a shift toward more-secure or otherwise politically preferred suppliers, or a supply chain designed to minimize environmental footprint.

For a youthful, poor country with a high fertility rate—one therefore expecting a surge of low-cost workers in the coming years—this could be a threat to export-driven development. But for middle-income countries that have already passed through their fattest demographic dividends and who now face slowing economic and workforce growth—places like Mexico, Brazil, Turkey, Thailand, Poland—these technology changes may actually offer new opportunity. Mexico will never be able to offer lower manufacturing wages than high-growth Egypt or Nigeria. But with the capacity to further substitute capital investment for labor, and the right trade and regulatory strategy, then it could continue to be competitive for given customer preferences.

Or consider the potential for technology to raise the quality and lower the cost of health care, as genome sequencing transforms biology into an information science, and machine-learning algorithms make assessments and recommendations drawing upon vast amounts of data from individual patients and from the electronic medical records and outcomes of millions of others. These rapidly advancing technologies portend quicker and more-accurate diagnoses, more-effective treatments, improved physician productivity, more-personalized medicine, new drugs and vaccines, early detection of pandemics, and much more, including an emphasis on prevention. Aging societies around the world, including our own, will see new demands for such capabilities. But it will not come automatically. Realizing these potential health benefits will require continued investment in research in the biological and information sciences, and another look at the way medical data is handled in order to allow scientists and engineers access to the great quantities of electronic medical-records data they need without compromising the privacy of individuals.

Other labor-saving emerging technologies can be used domestically to improve productivity in more-mature and demographically aging economies. Flying, sensor-equipped drones are being utilized for inspections of decaying

Japanese bridges and of fire-prone American forests, where they offer almost ten-to-one worker-hour advantages for otherwise dangerous work for which it is getting harder to find willing employees. Even China, which rode its own massive workforce bubble to great economic success over recent decades, is scrambling to apply artificial intelligence through its government and economy as, not yet rich, it already loses workers and stares down its "super-aged society" future. Chinese government investments in big data, machine learning, and image recognition have been framed as a contest with the United States for global hegemony. But they are also a response to a very real fear that without major productivity enhancements, its own consumption-based domestic economic growth is in peril.

This is part of the logic behind central-government mandates that local leaders act as public "anchor tenants"—effectively lead buyers—for privately developed artificial intelligence and data technologies to be incorporated into their towns and cities. Should the United States emulate China in this regard? Silicon Valley and other American academic clusters lead the world in advancing the science of many twenty-first-century technologies, including artificial intelligence. But near-term advances throughout the real economy (and national security), however, come from application of existing machine-learning technology to disrupt new fields and industries, a practice China has accelerated and in which other countries figure as well. Too much urgency may come at a cost, however. China's demographic and productivity challenge is greater than our own: unlike China, the United States can expect gradual, moderate workforce growth through this century across a diversified economy. We do not share the same crisis. And state-fueled investment in start-ups, or large purchase orders of the sorts of advanced technologies that might be favored by China's authoritarian government (like surveillance), can look like progress while actually distorting the overall market for these goods and services. So we think that the United States is right not to enter into an AI race with China, instead focusing on leading-edge technology plus quality, market-selected adoption in those applications where it is truly most valuable. The commercial market, not the federal government, is the best entity to allocate resources in this sector, as in most others. Should we wish to go faster with deployment, then subsidy, encouragement, or removal of regulatory roadblocks should flow through this same commercial system, with the government continuing to focus on funding the basic R&D that underpins broader advancement in these fields.

## Immigration and the Imperative of Governing over Diversity

We have observed how new technologies both allow people living in parts of the world with bad government, bad economies, or worsening environmental conditions to learn about the opportunities that exist elsewhere and facilitate their movement there. Young Syrian refugees in Germany chat daily with friends back home through mobile phones. The migrant caravans that now originate from Central America organize their journeys on Facebook.

Meanwhile, the demographic structures of many of the advanced, democratic countries to which global migrants wish to move are worsening, with low fertility rates, shrinking workforces (or even total populations), and fewer productive young people to support the elderly. This threatens the continued economic growth that has been taken for granted as the bedrock of modernity.

Put these two phenomena together and you should see mutual benefit, in theory. But the reality is the opposite, with countries around the world torn apart over this governance problem. The United States actually finds itself in better shape than many of its peers, in terms of both demographics and the politics of immigration (however unbelievable that claim may sound to those who tune in the national cable-news channels). Yet we too flounder. Previous "reasonable" efforts at US immigration reform over the past decade could not build the necessary political coalition to pass Congress and seem distant today. The McCain-Kennedy Secure America and Orderly Immigration Act of 2005, for example, included the idea of legalization and a path to citizenship for existing unauthorized immigrants, a guest worker program, and border security. Such an approach is commendable from a rule-of-law perspective, but one can also see where it stumbles: legalization is very attractive to current immigrants here illegally, but does little to address a wider range of concerns of the American electorate; guest worker programs similarly benefit those migrant workers and labor-needing employers, but perhaps equally turn off blue-collar US workers, who are already facing a variety of economic pressures and who worry that their labor pool will be diluted; this leaves border security which is basically agreed upon, if undefined in practice, and therefore finds itself a hostage to the rest of the deal.

We think that controlled migration—emphasizing the word *controlled*—will be good for the United States, and good for the vast majority of its citizens, across the hinge of history. Chancellor Angela Merkel's ongoing political debilitation in Germany, following her controversial 2015 decision to admit Syrian refugee numbers in excess of what many German citizens supported at the time, shows, however, that immigration policy is not solved by one interest temporarily overcoming another in the political arena. Floods and droughts will not work. So we suggest a new, longer-term approach that centers US citizen interests in the policy process as a way to build sustained support.

This outlook is influenced by our admittedly outlier perch in Stanford, California. "Harry" Lee Kuan Yew, the founder of modern Singapore, years ago visited the San Francisco Bay Area. Lee was here to find out about Stanford and Silicon Valley and how Singapore might learn from Americans to re-create such an atmosphere of innovation back home. His hosts agreed that it was a good idea but cautioned: "One thing you'll find, Harry, is that there are people here from all over the world." To this Lee replied, "I know that, but it could only happen in America." Here, immigration, much of it high skilled, is generally viewed favorably, because residents, both immigrant and native born, see it as positive for them and their communities. They know that the firms they work for or whose products and services they use were founded by or rely on workers who were not born here. They see themselves and their families as the cultural beneficiaries of that same window onto the world. All of which is cemented through daily, personal contacts with immigrants of all walks, humanizing what is otherwise a perilous philosophical question for modern democracies.

Is there a way to make immigration both appear to be, and substantively become, more positive for American host communities more broadly? That is, to address today's concerns and to demonstrate a narrative of value that will become increasingly apparent as the emerging world presents itself? In its goals, an immigration policy should deliver a sense of orderliness and safety, a feeling that migrants are bringing valued skills and attributes that match the needs of the local community without harming existing residents, and a sense of cultural compatibility. While immigration falls squarely in the camp of federal responsibilities, the failure of Washington, DC, to solve this problem has led to a cacophony of de facto local interventions; going forward, a sustained approach is likely to call more upon the insights of local government to structurally inform local needs and preferences. And while macro issues such as demographic balance and workforce needs are crucially important for the country, they also sound abstract. Consider how in Japan the general public shies away from open discussions of an immigration policy; at the same time, rural, small-town Japanese mayors take eager recruiting trips to Vietnam when they find themselves unable to fill their local kindergarten's bus driver positions. American citizens are also likely to have more confident preferences when they can be shown how these issues are playing out locally—or will play out—in their own communities, and how those lines may be connected to migration options.

The 2013 Bush-Bolick immigration proposals suggested a strategy along similar lines, with a preference for demand-based skilled immigration over today's family reunification policies to better match US labor market needs, as well as civics training and education to help new immigrants better understand the unique social and political foundations of the United States and more successfully assimilate into this system alongside existing citizens.[60] It is a place to start. We might further consider the degree to which the United States could try to better control the "push" of migrants it sees in the future, for example through interventions in likely or desired sending countries to affect the nature, education, and cultural knowledge of potential migrants. Again, the cost to do so would require sustained support by the American taxpayer—and as such would need to demonstrate concrete value to them.

To some degree, technology helps here in that it makes it easy for migrants to learn more about the culture, rights, and responsibilities of a potential receiving country while still at home. We have seen during our studies on Africa, for example, how technology facilitates the role of diasporas in effecting economic, social, and political changes in their former home communities. And their financial remittances are another underexamined organic resource through which source-country human capital investment can occur. Rising globally, total migrant remittances to low- and middle-income countries are now three times the level of foreign aid that those countries receive.[61] As the world's continued largest host of migrants, the United States has a variety of well-established and growing diasporas that could help spur such investments in their home countries.

## Maintaining National Security by Mastering Technology

Our national security also depends on taking advantage of revolutionary technologies while meeting the challenges posed by such technologies in the hands of adversaries. New low-cost systems such as autonomous drones that can swarm in large numbers, as well as small high-performance spacecraft, can give our forces additional capabilities and resilience. In-field additive manufacturing could simplify military logistics and improve readiness through just-in-time production and delivery of parts and materiel. Of course, they offer the same capability to any adversary who might master them.

We need to find the right mix of new technologies and improved versions of legacy systems, and the right posture for our forces. Collection and analysis of reliable information is increasingly critical—it's "the coin of the realm." In a world where cyberattacks on information networks are the norm, the side whose information most closely conforms to the real world is likely to prevail. Artificial intelligence will be necessary to analyze the vast amounts of data collected by modern sensors, but caution should be observed in using automated systems for decision making, as machines make mistakes. Great care should be taken to preserve the integrity of command-and-control networks, especially those in the nuclear arena.

COVID-19 has hit us like a ton of bricks. The risks were known, and we don't even need to go back to 1918 and the Spanish flu: the 1968 Hong Kong flu pandemic killed more Americans than the Vietnam and Korean wars did. This is a matter of national security, but it isn't the Department of Defense's war, and our governance institutions in 2020 were totally unprepared. The emerging new world creates the conditions that make future pandemics more possible. So we'll have to do better; we can, again, use concurrently emerging new technologies to help. Rapid diagnostics can help us identify new outbreaks quickly, monitor their spread, and understand their biology and their human or animal health impacts. Therapies and vaccines will be needed in short order and at scale, a dance in which researchers, private investors, and government regulators all have steps. Communication tools allow societies to engage in commerce and exchange even while apart. In fact, the most promising outcome we have seen so far in this calamity may well be just how adeptly those same private businesses and civil society members have harnessed these new technological capabilities in ways the federal government has not.

We forgot about risks to our global health, and we are paying for it. Similarly, we absolutely cannot forget—or, rather, we must re-remember—the unique danger of nuclear weapons, particularly ominous when viewed within the complex environment of high-end conventional weaponry and new competition in space and cyberspace. The circumstances in which nuclear weapons might be used seem to be growing, and further proliferation of nuclear weapons is a troublesome threat. Even a "regional" nuclear conflict would have serious global consequences. The United States and Russia, as the two largest nuclear powers, should preserve existing arms-control measures and work together to manage and reduce the existential threat to each other. Fortuitously, advances in big data, social media, and new technologies open up new opportunities for detecting nuclear-proliferation activities and monitoring arms-control agreements. And vast quantities of publicly available data can supplement traditional means of verifying international agreements. In short, we must embrace the challenges and opportunities that emerging technologies bring to our defense and war-fighting capabilities, because even if we do not, to recall again General Mattis, "the enemy gets a vote."

## Leading the World across the Hinge of History

The United States should be the country that can take the lead. But we aren't the leader we used to be. In fact, it is not just the United States that has lost this ability. No one today can go out to the world to command the respect that leaders such as Ronald Reagan, Helmut Kohl, Margaret Thatcher, and Yasuhiro Nakasone—and for that matter, truly transformative leaders including Deng Xiaoping and Mikhail Gorbachev—once did. Yet the emerging new world presents manifold opportunities that might rescue countries from seemingly intractable structural problems, be they of development or stagnation. And they will not be effectively harnessed on their own. In that sense, the stage is set for a broad defining of direction much as it was after the Second World War. The United States, recognizing the changes in store, and its own capabilities in navigating them, should strive to provide that direction.

This is a call not for the unilateral imposition of American ideas but for the United States to once again work with willing nations to set a better course. Postwar US leadership thrived by developing frameworks that benefited the United States, but that also allowed for the success of other states alongside it—other global powers have been unable to offer similarly compelling mutual frameworks. But leadership is a long-term investment, and US domestic enthusiasm for it has waned across both parties in the wake of the high costs and questionable results of military interventions, the chaotic international responses to the 2008 recession, poor public perception toward the fairness of and adherence to WTO agreements, and the apparent lack of tangible deliverables from talking bodies like the G7. Even bedrock security initiatives like NATO, crucial to the well-being of Americans and their allies, are held out as bargaining chips in the name of taxpayer burden. Each of these multilateral organizations has a purpose. They were created and led based upon the initiative of individuals who showed the American people what problems they would solve and what benefits they would offer. There are few such leaders today. Standing up multinational institutions, sustaining distant alliances, and supporting partners in more-violent regions are costly in terms of treasure and, sadly, blood. It is understandable that voters would question if the benefit is worth the cost. Whereas the likes of Harry Truman, Dean Acheson, George Marshall, and Dwight Eisenhower presented their case to the American people, elected leaders in recent years have failed to explain the benefit of or to make the case for continued American engagement abroad—and some have even denounced it. America has lost confidence in its own ability to lead, and the men and women in office have not restored it.

Does the emerging new world offer Americans a new reason to lead? Perhaps the failures of others to lead in our absence give a motivating sense of the world without US values at its center. Take the environment, where Europeans, embodied by Germany, have attempted grandiose efforts at global environmentalism, but have been unable to reach escape velocity from their own hypocritical domestic politics in doing so—their hectoring does not appear to scale, in stark departure from the successful reality-grounded approach embodied by the Montreal Protocol in the 1980s. Or information technology, where the French have reacted to their economy's own failure to produce successful internet products and services by throwing in the towel and promoting an adversarial tax regimen instead. At the other pole, China overtly offers its own tools of digital authoritarianism to aspiring autocrats who are similarly afraid of their own populace, or mobile-money financial technologies that deepen visibility into, and potential influence over, developing-economy consumers in Latin America and Africa. Even an emboldened Russia now sells modern weapons systems to splintering NATO allies.

These siloed initiatives do not live up to the challenges or opportunities of the emerging new world. No state or multilateral institution can fill the void of the United States. No other state has the free and open principles or the capacity needed to initiate international norms or consensuses. International organizations like NATO or the WTO lose their heft and legitimacy in the absence of active US participation, as do international agreements. Moreover, it is in America's interest, even narrowly defined, to preserve the principles of open seas, trade, and internet as well as the rule of law. International trade motivates economic growth, while the free exchange of ideas and technology makes people healthier and more comfortable. If all nations have access to the commons and are treated equally under the law, both Americans and citizens of every nation can benefit. These are principles worth sustaining, particularly as the race for new technologies quickens and global demographics rebalance. The challenge before us is to regain both the will and the capacity to do so.

The best way to get someone on your side isn't to try to convince them up front—rather, it's better to start working on some part of it and show good results. Global leadership can start at home, by getting our own house in order. Take that confidence in democracy, which has long been a US interest going beyond the borders of our own "experiment." But confidence seems to be waning globally as democratic politicians focus on eternal campaigning, which is a fundamentally divisive act, over the act of governing, which is about bridging differences. Worse: campaigning that reflects a personalization of politics to niche, motivated audiences almost as a form of personal political brand–building as opposed to broad statements of principles and strategies for the direction of the nation. Even in the United States, we see the declining effectiveness of Congress and the weakness of political parties as guiding institutions. The US political system has lost its way by veering away from its foundations as a representative democracy and toward increasing direct appeals—much of the personalization itself enabled by the communications tools of the emerging new world.

Structural reforms can help. The options to do so are well known. For example, returning power to committee chairs in Congress would reduce the authority of the leadership, which encourages polarization. Restoring structure to both chambers would also mitigate some of the media-friendly imbroglio driven by the most radical individual members. As would countering increasingly narrow, interest-based personalization by strengthening political parties themselves—perhaps by undoing McCain-Feingold Act campaign finance reforms, however well-intentioned they seemed at the time, which have resulted in a sidelining of these institutions.

Congress's ability to govern is likewise built on the US domestic fiscal situation. The antics that this institution now sustains itself with may be no surprise, given how little room it has actually left itself to tend to its main function, which is the purse. To sustain our technical, military, and international position, we will need to get our own fiscal house in order and free domestic resources for those priorities. Otherwise the US government will increasingly become, to borrow a phrase, nothing but "an insurance agency with a navy."

American leadership will depend in part on our ability to sustain our position at the forefront of the transformational technologies, and our military strength in the face of a wide range of challenges. Wise policies at home in these areas will substantially affect our ability to provide leadership abroad.

Personal leadership at home can make a difference. Contributors to our project have argued that when the United States faced its own "democratic distemper" in the 1960s and 1970s, it was the emergence of uncommonly strong leaders, here and among allies abroad, that turned society's tide by offering a positive and expansive vision. And none of the problems and issues identified here can be solved or addressed by the United States alone. We will need allies, partners, and friends. These are not in short supply if we pursue non-zero-sum policies. Most countries face the same challenges we do, in many cases more severe versions of them, and there is great potential for proceeding in partnership to address them, not only with our allies but with our competitors as well.

The spirit of our undertaking has been to understand the transformations brought on by rapidly changing demographics and advancing technology, and to develop strategies to seize the opportunities and mitigate the problems brought on by these unprecedented worldwide changes. Here are what we consider the key lessons for the United States in a world passing over a new hinge of history:

- Changing demographics pose profound challenges everywhere we look. Countries with young, growing populations need strong education systems, particularly for women and girls, and growing economies to foster reduced fertility rates and increased job opportunities. Countries with aging populations and shrinking workforces can look to new productivity-increasing technologies to help promote economic well-being across all age groups.
- Migration, managed properly with respect for the rule of law and the interests of all citizens, can benefit both those countries with young, growing populations and countries with aging populations. Given the projected growth, limited economic prospects, and disproportionate

impact of climate change in many developing countries, the United States should expect migration and prepare for it. One way to do that is to support education in developing countries, which has many benefits, including reduction in fertility, better job prospects at home, and providing those who migrate more to offer in their new locations.

- The United States benefits from migration. In contrast to many developed countries in Europe and Asia, our workforce continues to grow, thanks to migration. Our universities and our atmosphere of innovation in the commercial sector attract talented migrants from all over the world who contribute substantially to our technical prowess and economic success. Our experience with governing over diversity, while difficult, is one of our biggest assets going forward.
- The United States is also in a better position than most countries to take advantage of the benefits of advancing technology while mitigating the problems. Our companies and universities lead the commercial development of twenty-first-century technologies. We can look forward to increased productivity, a stronger economy, higher-quality health care at lower cost, and a better standard of living.
- Workers will need new skills to master new or redesigned jobs, and will have to change careers more frequently than in the past. Partnering community colleges with employers is a proven way to provide training to support transitions to new occupations. The social safety net can also be adapted to support these transitions.
- The biggest obstacle the United States faces is the shortfall in K–12 education. A good primary and secondary education is essential to providing everyone with a solid foundation for the lifelong learning and periodic reskilling necessary for success in the twenty-first-century workplace. Student performance should be the focus, rather than time in classroom, teacher seniority or credentials, or resources expended. Transparency, accountability, and flexibility should be central principles, based on accurate data on student performance. Every student deserves an education that provides an opportunity for success in life.
- Additive manufacturing and other advanced manufacturing techniques encourage making things near where they will be used and reduce incentives to source production in low-wage countries. Results include customized products (even unique products) and deglobalization, and

potential disruption of economies in countries pursuing manufacturing-led development.

- Advancing military technologies—precision conventional systems, drones, space and counterspace systems, offensive and defensive cyber capabilities—are transforming the nature of conflict. Our national security depends on taking advantage of these revolutionary technologies while meeting the challenges posed by such technologies in the hands of others. Collection and analysis of reliable information is increasingly critical ("the coin of the realm"). In a world where cyberattacks on information networks are the norm, the side whose information most closely corresponds to the real world is likely to prevail.
- Climate change is having an observable impact on the environment. Infectious diseases are moving north, and ecosystems and assets are being damaged. Policies to reduce emissions consistently, and on the necessary scale, include a revenue-neutral carbon tax and new investment in development and use of new technologies for clean energy and carbon removal.
- Nuclear weapons pose a unique threat, particularly ominous when viewed within the complex environment of advanced conventional systems and new competition in space and cyberspace. Further proliferation of nuclear weapons is a troublesome threat. Even a "regional" nuclear conflict would have serious global consequences.
- The United States is well positioned to take advantage of these global transformations—we have long experience, though still endeavoring, in governing over diversity, and our companies and universities lead the development and application of twenty-first-century technologies. We can face the future with confidence provided we take steps to prepare a process for welcoming migrants, reverse the decline in K–12 education, adapt our social safety net, protect our democratic processes, address vulnerabilities to cyber threats, and exploit the military applications of technology to our net benefit. In many cases, states can take the lead, serving as laboratories to develop innovative approaches.
- Our challenges can only be addressed through international cooperation. Traditionally, since the end of the Second World War, the United States has played a leadership role, encouraging the development of frameworks that benefited the United States but also allowed for the success of others. Our ability to resume a leadership role will depend

in part on sustaining our position at the forefront of the transformational technologies, sustaining our military strength in the face of a wide range of challenges, and getting our fiscal house in order. Wise policies at home will strengthen our ability to provide leadership abroad. We will need allies, partners, and friends—but most countries face the same challenges we do, and there is great potential for proceeding in partnership to address them, not only with our allies but with our competitors as well.

# Governance in an Emerging New World: Project Timeline

**October 3, 2018: Russia in an Emerging World**

Kori Schake, International Institute for Strategic Studies (moderator)
Ivan V. Danilin, Primakov National Research Institute of World Economy and International Relations
David Holloway, Freeman Spogli Institute
Min. Igor Ivanov, former minister of foreign affairs for Russia
Stephen Kotkin, Hoover Institution and Princeton University
Amb. Michael McFaul, Hoover Institution and Freeman Spogli Institute
Anatoly Vishnevsky, National Research University Higher School of Economics, Moscow

*(The three Russian papers were presented by Maria Smekalova, Russian International Affairs Council.)*

**October 29, 2018: China in an Emerging World**

Adm. Gary Roughead (USN, ret.), Hoover Institution (moderator)
Nicholas Eberstadt, American Enterprise Institute
Elsa Kania, Center for a New American Security and Harvard University
Kai-Fu Lee, Sinovation Ventures
Maria Repnikova, Georgia State University
Amb. Stapleton Roy, Wilson Center and former US ambassador to China

**November 13, 2018: The Information Challenge to Democracy**

Sec. Condoleezza Rice, Hoover Institution (moderator)
Niall Ferguson, Hoover Institution
Joseph Nye, Harvard University and the Hoover Institution

**December 3, 2018: Latin America in an Emerging World**

Min. Pedro Aspe, former secretary of finance of Mexico (moderator)
Min. Richard Aitkenhead, former minister of economics and minister of public finance of Guatemala
Silvia Giorguli, El Colegio de México
Claudia Masferrer León, El Colegio de México

Audience questions at the October 2018 panel on China in an Emerging New World. Hauck Auditorium, Stanford, California.

Ernesto Silva, Hoover Institution and former president of the Unión Demócrata Independiente (UDI) party in Chile
Ben Sywulka, Hapi.vc and Private Competitiveness Council of Guatemala

**January 14, 2019: Africa in an Emerging World**
Amb. George Moose, United States Institute of Peace and former assistant secretary of state for African affairs (moderator)
Tony Carroll, Manchester Trade
Amb. Chester A. Crocker, Georgetown University and former assistant secretary of state for African affairs
Mark Giordano, Georgetown University
Jack Goldstone, George Mason University
André Pienaar, C5 Capital

**February 4, 2019: Europe in an Emerging World**
Jim Hoagland, Hoover Institution and *Washington Post* (moderator)
Caroline Atkinson, former head of global policy for Google
Christopher Caldwell, *Claremont Review of Books*
William Drozdiak, Brookings Institution
Nicole Perlroth, *New York Times*
Jens Suedekum, Düsseldorf Institute for Competition Economics

Sec. Condoleezza Rice challenges historian Niall Ferguson on the impact of social networks at a public panel in November 2018.

**February 25, 2019: Emerging Technology and America's National Security**

Adm. James O. Ellis Jr. (USN, ret.), Hoover Institution (moderator)
Gen. Philip Breedlove (USAF, ret.), former Supreme Allied Commander Europe
Col. T. X. Hammes (USMC, ret.), National Defense University
Cpt. Katie Hedgecock, US Army and Stanford University
Margaret Kosal, Georgia Institute of Technology
Adm. Gary Roughead (USN, ret.), Hoover Institution and former chief of naval operations
Ralph Semmel, Johns Hopkins University Applied Physics Laboratory

**April 8, 2019: Health and the Changing Environment**

Lucy Shapiro, Stanford University and Beckman Center for Molecular & Genetic Medicine (moderator)
Harley McAdams, Stanford University
Dr. Kari Nadeau, Stanford University and Sean Parker Center for Allergy and Asthma Research
Stephen Quake, Stanford University and Chan Zuckerberg Biohub
Dr. Milana Boukhman Trounce, Stanford Emergency Medicine, Stanford BioSecurity and Infectious Disease Disaster Response

Political scientist Joseph Nye in conversation with Stanford students following a November 2018 discussion of international engagement norms in Hoover's David and Joan Traitel Building pavilion.

**April 22, 2019: The Middle East in an Emerging World**

Abbas Milani, Hoover Institution and Stanford University (moderator)
Prince Moulay Hicham Alaoui, Harvard University
Houssem Aoudi, Afkar Tunisia
Lisa Blaydes, Stanford University
Arye Carmon, Hoover Institution and Israel Democracy Institute
Aykan Erdemir, Foundation for Defense of Democracies
Roya Pakzad, Stanford University

**May 6, 2019: Emerging Technology and the US Economy**

Gopi Shah Goda, Stanford Institute for Economic Policy Research (moderator)
Erik Brynjolfsson, MIT Sloan School and MIT Initiative on the Digital Economy
Dipayan Ghosh, Shorenstein Center at Harvard University and former White House adviser
Jim Hollifield, Southern Methodist University and Tower Center
John B. Taylor, Hoover Institution and Stanford University
Van Ton-Quinlivan, California Community Colleges

Former governor of Florida Jeb Bush on the challenges facing elected office holders in an information age in an October 2019 project session. (Patrick Beaudouin for the Hoover Institution)

**May 14, 2019: Stability in an Age of Disruption**

Deborah Gordon, Stanford Preventive Defense Project (moderator)

Larry Diamond, Hoover Institution, Freeman Spogli Institute, and Stanford Global Digital Policy Incubator

Mo Fiorina, Hoover Institution and Stanford University

Jack Goldstone, Center for the Study of Social Change, Institutions, and Policy, George Mason University

Judge Alice Hill, Hoover Institution and former member of the US National Security Council and Department of Homeland Security

Charles Hill, Hoover Institution and Yale University, former US State Department and United Nations special adviser

**September 10, 2019: Roundtable on Japan**

Amb. Michael Armacost, Freeman Spogli Institute and former US ambassador to Japan (moderator)

Michael R. Auslin, Hoover Institution

Karen Eggleston, Freeman Spogli Institute

Kenji Kushida, Freeman Spogli Institute and Silicon Valley–New Japan Project

George Shultz moderates an October 2019 roundtable on cyber weapons in the Hoover Institution's Annenberg Conference Room.

**September 17, 2019: Roundtable on K–12 Education**
Eric A. Hanushek, Hoover Institution (moderator)
Adam Carter, Summit Charter Schools
Margaret (Macke) Raymond, Hoover Institution
Christopher N. Ruszkowski, Hoover Institution

**September 19, 2019: Roundtable on Bangladesh**
Sanchita Saxena, UC Berkeley Institute for South Asian Studies

**October 7, 2019: Governing in an Emerging New World**
Jim Hoagland, Hoover Institution and *Washington Post* (moderator)
Willie Brown, former mayor of San Francisco and speaker of the California State Assembly
Gov. Jeb Bush, former governor of Florida
Amanda Daflos, chief innovation officer for the mayor of Los Angeles
Chris DeMuth, Hudson Institute
Dan Henninger, *Wall Street Journal*
Karen Tumulty, *Washington Post*

Hoover senior fellows Amy Zegart and Larry Diamond debate the role of deterrence in cyber warfare.

**October 29, 2019: Roundtable on Cybertech**

Amy Zegart, Hoover Institution and Freeman Spogli Institute (moderator)
Herbert Lin, Hoover Institution and Freeman Spogli Institute
Samantha Ravich, Foundation for Defense of Democracies
Jacquelyn Schneider, Hoover Institution

**November 5, 2019: Emerging Technology and Nuclear Nonproliferation**

Elisabeth Paté-Cornell, Stanford University and former member of the President's Foreign Intelligence Advisory Board (moderator)
Sec. Ernest J. Moniz, Nuclear Threat Initiative and former US secretary of energy
Sen. Sam Nunn, Nuclear Threat Initiative, Hoover Institution, former US senator, and chairman of the Senate Armed Services Committee
Sec. William J. Perry, Hoover Institution and former US secretary of defense
Ashley Tellis, Carnegie Endowment for International Peace and former senior advisor at the US embassy in New Delhi

George Shultz looks on during an October 2019 project panel discussion for the local Stanford and Silicon Valley community. (Patrick Beaudouin for the Hoover Institution)

**November 20, 2019: Roundtable on India**
Amb. David C. Mulford, Hoover Institution and former US ambassador to India (moderator)
Adele M. Hayutin, Hoover Institution
Dinsha Mistree, Stanford Law School

**December 12, 2019: Concluding Roundtable**
Sec. George P. Shultz, Hoover Institution (moderator)

*Note: While the authors of this book benefited immensely from the insights of these participants and many others throughout the project, the views in this book are not endorsed by the discussants listed here or their organizations. Affiliations are listed for identification only and are accurate as of the time of the indicated session.*

# Notes

1. Our World in Data, a project of the UK-based Global Change Data Lab, is an excellent source of long-term statistics of these and other major global human progress indicators. See, for example, Max Roser, "The Short History of Global Living Conditions and Why It Matters That We Know It," *Our World in Data* (2019), https://ourworldindata.org/a-history-of-global-living-conditions-in-5-charts.
2. Stanley Lebergott, "Labor Force and Employment, 1800–1960" in *Output, Employment, and Productivity in the United States after 1800,* edited by Dorothy S. Brady, National Bureau of Economic Research, 1966, http://www.nber.org/chapters/c1567.
3. Bureau of Labor Statistics, "Job Openings and Labor Turnover Survey," June 2019, https://www.bls.gov/jlt.
4. Bangladesh Bureau of Statistics, "Labor Force Survey 2017" (March 2019), http://www.bbs.gov.bd/site/page/111d09ce-718a-4ae6-8188-f7d938ada348/-; and Rachel Heath and A. Mushfiq Mobarak, "Manufacturing Growth and the Lives of Bangladeshi Women," National Bureau of Economic Research, NBER Working Paper 20383 (August 2014), http://www.nber.org/papers/w20383.
5. Bangladesh Garment Manufacturers and Exporters Association, "Trade Information," website accessed February 2020, http://www.bgmea.com.bd/home/pages/TradeInformation.
6. Nguyen Thi Lan Huong, "Vietnamese Textile and Apparel Industry in the Context of FTA," presented March 16, 2017, to United Nations Economic and Social Commission for Asia and the Pacific (UNESCAP) meeting in Hanoi, https://www.unescap.org/sites/default/files/DA9%20Viet%20Nam%20Session%207%20-%20textile%20and%20apparel%20industry.pdf.
7. International Federation of Robotics, "World Robotics 2018: Industrial Robots" (2018), https://ifr.org/downloads/press2018/Executive_Summary_WR_2018_Industrial_Robots.pdf.
8. Jeremiah Dittmar, "The Welfare Impact of New Good: The Printed Book," working paper, February 2012.
9. Adele Hayutin is the author of a forthcoming book from Hoover Institution Press on the topic of global demographics and their social, economic, and governance implications.
10. These and other projections discussed here are from the United Nations, *World Population Prospects: The 2019 Revision*, medium variant. Prepared by

the population division of the department of economic and social affairs of the United Nations secretariat, https://population.un.org/wpp.

11. International Organization for Migration, *World Migration Report 2020*, Migration Policy Research Division, https://publications.iom.int/books/world-migration-report-2020.
12. Stuart Anderson, "Immigrants and Billion-Dollar Companies," National Foundation for American Policy (October 2018), https://nfap.com/wp-content/uploads/2016/03/Immigrants-and-Billion-Dollar-Startups.NFAP-Policy-Brief.March-2016.pdf.
13. Chang-Tai Hsieh and Enrico Moretti, "Housing Constraints and Spatial Misallocation," National Bureau of Economic Research, NBER Working Paper 21154, May 2017.
14. Eric A. Hanushek, Paul E. Peterson, Laura M. Talpey, and Ludger Woessmann, "The Achievement Gap Fails to Close," *Education Next* 19, no. 3 (Summer 2019), https://www.educationnext.org/achievement-gap-fails-close-half-century-testing-shows-persistent-divide.
15. "Remarks by the President in Discussion with Ninth Graders" delivered by President Barack Obama at Wakefield High School, Arlington, Virginia, on September 8, 2009, https://obamawhitehouse.archives.gov/realitycheck/the-press-office/remarks-president-discussion-with-9th-graders-wakefield-high-school.
16. Daniel DiSalvo, "Teachers Want Higher Pay, but Pensions Swallow Up the Money," *Wall Street Journal*, September 30, 2019, https://www.wsj.com/articles/teachers-want-higher-pay-but-pensions-swallow-up-the-money-11569885425.
17. Cory Koedel, "California's Pensions Debt Is Harming Teachers and Students Now—and It's Going to Get Worse," *Brown Center Chalkboard*, Brookings Institution, May 3, 2019, https://www.brookings.edu/blog/brown-center-chalkboard/2019/05/03/californias-pension-debt-is-harming-teachers-and-students-now-and-its-going-to-get-worse.
18. Walter Russell Mead, "The Big Shift: How American Democracy Fails Its Way to Success," *Foreign Affairs* 97, no. 3 (May/June 2018).
19. Carl Benedikt Frey and Michael A. Osborne, "The Future of Employment: How Susceptible Are Jobs to Computerisation?" Oxford Martin programme on technology and employment working paper, September 2013, https://www.oxfordmartin.ox.ac.uk/downloads/academic/The_Future_of_Employment.pdf.
20. Melanie Arntz, Terry Gregory, and Ulrich Zierahn, "The Risk of Automation for Jobs in OECD Countries," *OECD Social, Employment and Migration Working Papers*, May 14, 2016, https://doi.org/10.1787/5jlz9h56dvq7-en.
21. Organisation for Economic Co-Operation and Development (OECD), "The Future of Work," *OECD Employment Outlook 2019*, April 25, 2019, https://doi.org/10.1787/9ee00155-en.
22. McKinsey Global Institute, "Jobs Lost, Jobs Gained: Workforce Transitions in a Time of Global Automation," November 28, 2017,

https://www.mckinsey.com/featured-insights/future-of-work/jobs-lost-jobs-gained-what-the-future-of-work-will-mean-for-jobs-skills-and-wages.

23. Bureau of Labor Statistics, "Number of Jobs, Labor Market Experience, and Earnings Growth: Results from a National Longitudinal Survey," April 1, 2015, https://www.bls.gov/news.release/archives/nlsoy_03312015.htm.
24. Signe-Mary McKernan, Caroline Ratcliffe, and Stephanie R. Cellini, "Transitioning In and Out of Poverty," Urban Institute, Fact Sheet No. 1 (September 2009), https://www.urban.org/sites/default/files/publication/30636/411956-Transitioning-In-and-Out-of-Poverty.PDF.
25. George P. Shultz and George B. Baldwin, "Automation: A New Dimension to Old Problems," *Annals of American Economics* (Washington: Public Affairs Press, 1955).
26. Michael D. Bordo and John V. Duca, "The Impact of the Dodd-Frank Act on Small Business," Hoover Institution Economics Working Paper 18106, April 2, 2018, https://www.hoover.org/sites/default/files/research/docs/18106-bordo-duca.pdf.
27. Daniel Henninger, "Can the West Still Govern?," *Wall Street Journal*, June 5, 2019, https://www.wsj.com/articles/can-the-west-still-govern-11559774606.
28. Christopher DeMuth, "Repairing Our Fractured Politics" (remarks at the American Enterprise Institute, December 14, 2017), https://www.hudson.org/research/14106-repairing-our-fractured-politics.
29. Christopher DeMuth, "Trumpism, Nationalism, and Conservatism," *Claremont Review of Books*, vol. 19, no. 1 (Winter 2019), https://www.claremontreviewofbooks.com/trumpism-nationalism-and-conservatism.
30. David Pierson and Makeda Easter, "Lyft Is Turning Uber's Missteps into an Opportunity," *Los Angeles Times*, June 13, 2017.
31. Calculated using data from the United Nations Population Division's *2017 Revision of World Population Prospects*, available at https://population.un.org/wpp.
32. Calculated using data from the United Nations, Department of Economic and Social Affairs, "World Population Prospects 2019," available at https://population.un.org/wpp.
33. US Census Bureau, "Driving Population Growth: Projected Number of People Added to U.S. Population by Natural Increase and Net International Migration," March 13, 2018, https://census.gov/library/visualizations/2018/comm/international-migration.html.
34. World Bank Gender Equality Indicators, "Fertility rate, total (births per woman)," retrieved from https://data.worldbank.org/indicator/SP.DYN.TFRT.IN.
35. Calculated using data from the United Nations, "World Population Prospects 2019."
36. National Center for Health Statistics, *Health, United States, 2017: With Special Feature on Mortality* (Hyattsville, MD: Centers for Disease Control and Prevention, 2018), https://www.cdc.gov/nchs/hus/index.htm.

37. Jynnah Radford, "Key Findings about U.S. Immigrants," Pew Research Center publication, June 17, 2019, https://www.pewresearch.org/fact-tank/2019/06/17/key-findings-about-u-s-immigrants.
38. Noah Smith, "Workers at the Bottom Are the Last to Gain," *Bloomberg Opinion*, June 17, 2019, https://www.bloomberg.com/opinion/articles/2019-06-17/workers-at-the-bottom-are-last-to-gain-in-economic-recoveries.
39. James Madison, "The Same Subject Continued: The Union as a Safeguard against Domestic Faction and Insurrection," *Federalist*, no. 10, (November 1787) as reprinted in the Library of Congress Online Research Guide, https://guides.loc.gov/federalist-papers/full-text.
40. A number of analysts have argued that the inclusion of PM2.5 fine particulate matter pollution, a commonly used if somewhat less universally accepted measure, would drive this number to more than 100,000 deaths annually. See, for example, Andrew L. Goodkind, Christopher W. Tessum, Jay S. Coggins, Jason D. Hill, and Julian D. Marshall, "Fine-Scale Damage Estimates of Particulate Matter Air Pollution Reveal Opportunities for Location-Specific Mitigation of Emissions," *Proceedings of the National Academy of Sciences* 116, no. 18 (April 30, 2019): 8775–80, doi.org/10.1073/pnas.1816102116. For an earlier estimate, see Fabio Caiazzo, Akshay Ashok, Ian A. Waitz, Steve H. L. Yim, and Steven R. H. Barrett, "Air Pollution and Early Deaths in the United States. Part I: Quantifying the Impact of Major Sectors in 2005," *Atmospheric Environment* 79 (November 2013): 198–208, doi.org/10.1016/j.atmosenv.2013.05.081.
41. Kathryn Olivarius, "Immunity, Capital, and Power in Antebellum New Orleans," *American Historical Review* 124, no. 2 (April 2019): 425–55, doi.org/10.1093/ahr/rhz176.
42. Centers for Disease Control, "Elimination of Malaria in the United States (1947–1951)," July 23, 2018, https://www.cdc.gov/malaria/about/history/elimination_us.html.
43. Centers for Disease Control, "National Vital Statistic System HIST290: Historical Age-Specific Death Rates," November 6, 2015, https://www.cdc.gov/nchs/nvss/mortality/hist290.htm.
44. Charbel el Bcheraoui, Ali H. Mokdad, Laura Dwyer-Lindgren, Amelia Bertozzi-Villa, Rebecca W. Stubbs, Chloe Morozoff, Shreya Shirude, Mohsen Naghavi, and Christopher J. L. Murray, "Trends and Patterns of Differences in Infectious Disease Mortality Among US Counties, 1980–2014," *Journal of the American Medical Association* 319, no. 12 (March 27, 2018): 1248–60, doi:10.1001/jama.2018.2089.
45. Yacov Zahavi, "Equilibrium Between Travel Demand System Supply and Urban Structure," in Transport Decisions in an Age of Uncertainty: The Proceedings of the Third World Conference on Transport Research, Rotterdam, 26–28 April, 1977, ed. Evert Visser (The Hague: Nijhoff, 1977).
46. US Environmental Protection Agency, "Air Pollutant Emissions Trends Data: National Annual Emissions Trend—Criteria Pollutants National Tier 1 for

1970–2019," updated April 27, 2020, https://www.epa.gov/air-emissions-inventories/air-pollutant-emissions-trends-data.

47. Clean Air Task Force, "Toll From Coal Interactive Map," 2018, http://www.catf.us/educational/coal-plant-pollution.
48. National Academies of Sciences, Engineering, and Medicine, *A Research Review of Interventions to Increase the Persistence and Resilience of Coral Reefs* (Washington, DC: The National Academies Press, 2019), https://doi.org/10.17226/25279.
49. Ashish Sharmal, Conrad Wasko, and Dennis P. Lettenmaier, "If Precipitation Extremes Are Increasing, Why Aren't Floods?," *Water Resources Research* 54, no. 11 (November 2018): 8545–51, https://doi.org/10.1029/2018WR023749.
50. Harold E. Brooks and Charles A. Doswell III, "Deaths in the 3 May 1999 Oklahoma City Tornado from a Historical Perspective," *American Meteorological Society* 17, no. 3 (July 2001): 354–61.
51. US Department of Agriculture, "National Agriculture Statistics Service: Quick Stats," data for 2019, https://quickstats.nass.usda.gov.
52. US Energy Information Administration, "Fort Calhoun Becomes Fifth U.S. Nuclear Plant to Retire in Past Five Years," *Today in Energy*, October 31, 2016, https://www.eia.gov/todayinenergy/detail.php?id=28572.
53. Patrick Monreal, "Inside Stanford's Last Fallout Shelter: A Time Capsule to Cold War Politics and Protests," *Stanford Daily*, September 15, 2019, https://www.stanforddaily.com/2019/09/25/inside-stanfords-last-fallout-shelter.
54. Owen B. Toon, Charles G. Bardeen, Alan Robock, Lili Xia, Hans Kristensen, Matthew McKinzie, R. J. Peterson, Cheryl S. Harrison, Nicole S. Lovenduski, and Richard P. Turco, "Rapidly Expanding Nuclear Arsenals in Pakistan and India Portend Regional and Global Catastrophe," *Science Advances* 5, no. 10 (October 2019), eaay5478.
55. Toon et al., "Nuclear Arsenals"; Jon Reisner, Gennaro D'Angelo, Eunmo Koo, Wesley Even, Matthew Hecht, Elizabeth Hunke, Darin Comeau, Randall Bos, and James Cooley, "Climate Impact of a Regional Nuclear Weapons Exchange: An Improved Assessment Based on Detailed Source Calculations," *Journal of Geophysical Research: Atmospheres* 123, no. 5 (March 16, 2018), 2752–72; Raymond Jeanloz, "Environmental Effects of Nuclear War," in *Andrei Sakharov: The Conscience of Humanity*, eds. Sidney D. Drell and George P. Shultz (Stanford, CA: Hoover Institution Press, 2015): 53–68.
56. US Office of Technology Assessment, *The Effects of Nuclear War* (Washington, DC: US Government Printing Office, 1979), https://ota.fas.org/reports/7906.pdf; Princeton Science & Global Security, "PLAN A," internet video, 4:18, posted September 2019, https://sgs.princeton.edu/the-lab/plan-a.
57. Joshua Coupe, Charles Bardeen, Alan Robock, and Owen Toon, "Nuclear Winter Responses to Nuclear War Between the United States and Russia in the Whole Atmosphere Community Climate Model Version 4 and the Goddard

Institute for Space Studies ModelE," *Journal of Geophysical Research: Atmospheres* 124, no. 15 (August 16, 2019): 8522–43.

58. William J. Perry, "Bill Perry's D.C. Nuclear Nightmare," The William J. Perry Project. YouTube video, 5:40, posted March 2016, https://www.wjperryproject.org/nuclear-nightmares.
59. Kenji Morita, "Meiji-Era Advisors," *Japanology Plus,* NHK Broadcasting television program, December 11, 2018.
60. As outlined in Jeb Bush and Clint Bolick, *Immigration Wars: Forging an American Solution* (New York: Simon & Schuster, 2013).
61. Phillip Connor, D'Vera Cohn, and Ana Gonzalez-Barrerra, "Changing Patterns of Global Migration and Remittances," Pew Research Center publication, December 17, 2013, https://www.pewsocialtrends.org/2013/12/17/changing-patterns-of-global-migration-and-remittances.

# About the Authors

**George Pratt Shultz** is the Thomas W. and Susan B. Ford Distinguished Fellow at the Hoover Institution. He has had a distinguished career in government, academia, and the world of business. He is one of two individuals who have held four different federal cabinet posts; he has taught at three of this country's great universities; and for eight years he was president of a major engineering and construction company. He attended Princeton University, graduating with a BA in economics, whereupon he enlisted in the US Marine Corps, serving in the Pacific from 1942 through 1945. He later earned a PhD in industrial economics from the Massachusetts Institute of Technology and served on President Eisenhower's Council of Economic Advisers. Shultz was dean of the University of Chicago Booth School of Business from 1962 to 1969, before returning to Washington to serve as secretary of labor, then as director of the Office of Management and Budget and as secretary of the Treasury, in the cabinet of President Nixon. Shultz was sworn in July 16, 1982, as the sixtieth US secretary of state under President Reagan and served until January 20, 1989. In 1989, Shultz was awarded the Medal of Freedom, the nation's highest civilian honor. Among recent publications, he was editor of *Blueprint for America* (2016), author of *Thinking about the Future* (2019), and with John B. Taylor coauthor of *Choose Economic Freedom* (2020).

**James Timbie** is an Annenberg Distinguished Visiting Fellow at the Hoover Institution and, along with George P. Shultz, co-convener of the Governance in an Emerging New World project. As a senior adviser in the State Department from 1983 to 2016, Timbie played a central role in the negotiation of the INF and START nuclear arms–reduction agreements, the purchase from Russia of enriched uranium extracted from dismantled nuclear weapons for use as fuel to produce electricity in the United States, and the establishment of an international enriched uranium fuel bank. More recently, he was the lead US expert in the negotiation of the nuclear agreement with Iran. He retired from the State Department in 2016 and in 2018 was coeditor with George P. Shultz and Jim Hoagland of *Beyond Disruption: Technology's Challenge to Governance.* A Princeton graduate, he has a PhD in physics from Stanford University and from 1971 to 1983 was a scientist at the Arms Control and Disarmament Agency.

# Index